Prep for the AP Exam Guide
for

Yates, Moore, and Starnes's

THE PRACTICE
OF STATISTICS
Second Edition

TI-83/89 Graphing Calculator Enhanced

Larry J. Peterson

W. H. Freeman and Company
New York

D1085938

Printed in the United States of America

ISBN: 0-7167-9615-5

First printing 2002

CONTENTS

PREFACE

Prep for Exam Supplement is one of several supplements to *The Practice of Statistics,* Second Edition, written by Daniel Yates, David Moore, and Daren Starnes. This book has been prepared with one over-riding purpose: to help students prepare for the Advanced Placement Statistics Exam. In 1996 the first examination was given for AP Statistics. Nearly 7500 students took the first exam. Since that time, the number of students taking AP Statistics has grown beyond 50,000.

AP Statistics has been an exciting and rewarding course to teach. A former student taught me the essence of this course when he said, "AP Statistics has made me a better consumer of information."

Many excellent supplements have been published to assist students in their preparation for the exam. This supplement is unique as it is cross-referenced to the text: *The Practice of Statistics,* Second Edition. Each practice exam has a complete answer key that explains the theory and/or computations for each question.

The supplement consists of four full-length practice exams with 40 multiple-choice questions and six free-response questions. Teachers can assign all or part of an exam to help students review the course. Students will benefit by taking the practice exams under conditions that simulate the actual testing environment. While technology is useful in solving the questions, the emphasis is on understanding and communication of the concepts

The questions are in random order to stimulate a full range of thinking that will review all the concepts and techniques listed in the Advanced Placement Statistics Course Description published by The College Board. By completing all practice examinations, students can discover their strengths and weaknesses. They can then approach the actual AP Exam with increased confidence.

ACKNOWLEDGMENTS

I want to thank Mark Santee and Danielle Swearengin from W.H. Freeman for their excellent guidance in keeping this work on schedule and for their suggestions to improve the manuscript.

I appreciate the confidence that Dan Yates and Daren Starnes showed in me by recommending me to W.H. Freeman to write this supplement. They have offered many thoughtful and insightful suggestions and corrections to my original work. I have grown professionally as a teacher and as an author as a result.

I would like to thank my parents, John and Cleo Peterson, whose deafness never kept them from enjoying life and showing me that life is good. Finally, to my wife, Lynette, thank you for supporting me and encouraging me in my career as a math teacher; a career that has brought me incredible joy and satisfaction.

Practice Exam 1

#1   A concert hall has 2000 seats. There are 1200 seats on the main floor and 800 in the balcony. 40% of those in the balcony buy a souvenir program. 50% of those on the main floor buy a souvenir program. At a certain performance all seats are occupied. If an audience member is selected at random, what is the probability that a program was purchased?

(A)   22.5%
(B)   44%
(C)   45%
(D)   46%
(E)   92%

#2   Researchers conducted a 23-year study among 10,125 government employees in a certain country. All were men over the age of 40, and eventually 369 died of strokes. Participants were surveyed at the beginning about their perception of family problems. Among the 9000 who reported a happy personal life, 3.5% died from strokes. Among the 1125 who reported serious problems in their personal life, the death rate from strokes was 4.8%. What is the P-value of the appropriate test to see if there is a relationship between personal happiness and death from strokes?

(A)   0.028
(B)   0.031
(C)   0.042
(D)   0.062
(E)   0.083

#3   A car dealer wants to estimate the proportion of pickup trucks in a large city compared to all vehicles. He decides to take several random samples of size 100 from the parking lot at a baseball stadium using the numbers that are already assigned to each stall. Which of the following is NOT a correct statement?

(A)   Because of sampling variability, each sample could produce a different proportion.
(B)   The margin of error for a 90% confidence interval is independent of the number of trucks in the parking lot.
(C)   The sample may be biased since baseball fans may not have the same preferences as the general population.
(D)   If a particular numbered stall selected at random does not have a truck, it is acceptable to continue examining stalls to the left until a stall with a truck is found.
(E)   It is improper to select the first 100 stalls closest to the main gate since those owners are likely to have reserved parking, which is an indication of income levels.

#4   The mean score for a round of golf by Tiger Woods during the 1999 and 2000 seasons was 68.04 with a standard deviation of 2.4. Assuming that his scores are independent and normally distributed, what is the probability that he will shoot two consecutive scores of 66 or less?

(A)   0.04
(B)   0.08
(C)   0.11
(D)   0.20
(E)   0.39

#5   A spinner has an equal chance of landing on any one of the numbers 1-10. A student spins the spinner 100 times and finds the total of the numbers. This is repeated 150 times. The results of each trial are plotted in a histogram. Which of the following best describes the shape of the histogram?

(A)   Normally distributed with a mean of 550.
(B)   Binomially distributed with a mean of 550.
(C)   Uniformly distributed with a mean of 550.
(D)   Continuously distributed with a mean of 550.
(E)   Cannot be determined since the mean of the sample was not computed.

#6   Doctors believe that patients who received heart pacemakers seemed to snore less. Among 40 randomly selected patients with a pacemaker, 12 snored. Among 60 randomly selected patients without a pacemaker, 25 snored. Which of the following statements is NOT correct?
I.   The evidence supports the theory at the 10% level
II.   The 95% confidence interval for the difference between the proportion of the two groups that snore is (-0.31, 0.07)
III.   Since the confidence interval contains 0, there is a statistical difference between the two proportions.

(A)   I only
(B)   II only
(C)   III only
(D)   I and III
(E)   I, II, and III are incorrect

#7   A strength coach wants to determine if there is a relationship between the weight of an individual and how much weight the individual can lift. The data was collected and analyzed using a statistical software program. The output is shown below.

| TotalWeightLifted | StudentWeight | | Total |
	Under 150	Over 150	
Under 200	44 40.0 0.40	43 47.0 0.34	87 87.0 0.00
Over 200	25 29.0 0.55	38 34.0 0.47	63 63.0 0.00
Total	69 69.0	81 81.0	150 150.0

The number of rows with at least one missing value is 0

Chi-Square Statistics Section

Chi-square	1.745200
Degrees of freedom	1.0000
Probability level	0.18641
Phi	.107864
Cramer's V	.107864
Kappa's t-value	1.32106
McNemar's Probability Level	.029049

Based on the information, which of the following is a correct statement?

(A)   Since the value of the χ^2 statistic is so small, there is a strong relationship between the weight of an individual and the weight the individual can lift.

(B)   Since the value of the χ^2 statistic is so small, there is no relationship between the weight of an individual and the weight the individual can lift.

(C)   The correlation coefficient of a linear regression equation would be less than 0.20.

(D)   The P-value of this test is 0.18641 which indicates a weak relationship between the weight of an individual and the weight the individual can lift.

(E)   The P-value of this test is 0.029049 which indicates a very strong relationship between the weight of an individual and the weight the individual can lift.

#8   The following is a stem and leaf plot for the number of defective chips in a batch of 300 chips for two different machines.

Machine A		Machine B
	1	1
99888	1	57778899
443333222222	2	001124444
8888777666666655555	2	66777889
1100	3	00001123
	3	6688
	4	0
	4	
	5	1
		2|1 means 21

Based on the plot, which of the following statements is TRUE?
I.   The output from Machine B has more variability than that of Machine A.
II.   The mean from Machine A is greater than the mean from Machine B.
III. The data from Machine A has less skewness than from Machine B.

(A)   I only
(B)   II only
(C)   III only
(D)   I and III only
(E)   All statements are true

#9   An author wrote a book that advocated the playing of classical music to young children. The author cited a study that reported that students who choose to play a musical instrument score 51 points higher on the Scholastic Aptitude Test than students who do not play an instrument. Which of the following statements is NOT correct?
I.   There is a relationship between playing musical instruments and scoring higher on the SAT.
II.   The study was observational, rather than experimental.
III. Playing a musical instrument will cause a student to score higher on the SAT.
IV.   The study was a matched pairs study.

(A)   I and II
(B)   II and III
(C)   I and IV
(D)   II and IV
(E)   III and IV

6

#10   Randomization is an important characteristic of a well-designed experiment in order to eliminate bias. Which of the following sources of bias is eliminated by randomization?

(A)   Researchers have a desired outcome in mind.
(B)   Certain groups are systematically excluded from the data.
(C)   Certain groups are categorically excluded from the data.
(D)   People with strong opinions are counted before others.
(E)   Some groups are aware of the reason for the data search.

#11   A spinner on a full circle can take on any decimal value between 0 and 400. What is the probability that the spinner will land between 175 and 225?

(A)   0.12
(B)   0.14
(C)   0.44
(D)   0.50
(E)   0.56

#12   A poll of a random sample of 1068 people in Sweden (total population of 9 million) found that about 75% favored government supported health care. A similar poll of 1068 people from England (population 60 million) showed that about 75% favored government supported health care. Which of the following statements is TRUE?

(A)   Since the same number of people had the same opinions, the results are equally accurate.
(B)   Since the Swedish poll represented a larger proportion of the population, it is more accurate.
(C)   Since the population of England is larger, the English poll gives a better indication of the opinions of the people.
(D)   No decision can be made since the confidence interval is not stated.
(E)   No decision can be made since the populations are different and the standard deviations are unknown.

#13   According to the U.S. Census, the proportion of adults in a certain county who owned their own home was 0.71. An SRS of 100 adults in a certain section of the county found 65 owned their homes. What is the probability of obtaining a sample of 100 adults in which 65 or fewer own their own homes, assuming that this section of the county has the same overall proportion of adults who own their homes as does the entire county?

(A)   0
(B)   0.093
(C)   0.106
(D)   0.186
(E)   0.907

#14   After a snowstorm in a large metropolitan area, meteorologists took a random sample of several locations and measured depth of the snow along with the water content. The results were summarized in a computer printout.

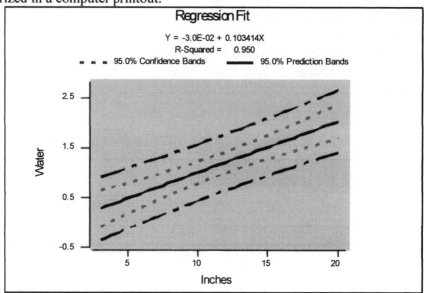

LINEAR REGRESSION ANALYSIS
The regression equation is
Water = -0.03039 + 0.10341*Inches
R SQR = 0.95020
R = 0.97478
DF = 4
T = 8.610
P = ########

Unfortunately, the printer failed just as the P-value was being displayed. What is the significance level for this hypothesis test H_0: $\beta = 0$ vs. H_1: $\beta \neq 0$?

(A)   P < 0.001
(B)   P = 0.001
(C)   P = 0.0304
(D)   P = 0.103
(E)   P = 0.950

#15   The monthly power usage in a rural area is given by $\hat{y} = 2.7 + 1.01x$ where $\hat{y}$ is the number of kilowatts used and x represents the number of months since January 1, 2000 (Let January 2000 correspond to $x = 0$). What is the estimated monthly power usage for December 2002?

(A)   5.73 kw
(B)   25.93 kw
(C)   26.94 kw
(D)   38.05 kw
(E)   39.06 kw

8

#16   Eight people who suffer from hay fever volunteer to test a new medication that will relieve the symptoms. The names of the volunteers are

1.   Zavala   5.   Harris
2.   Liu   6.   Elifritz
3.   Lonnquist   7.   Klein
4.   McGarry   8.   Scott

Four of the volunteers will receive the new medication while the other four will receive a placebo as part of a double blind experiment. Starting at the left of the list of random numbers below and reading from left to right, assign four people to be given the medication.

07119   97336   71048   08178   77233   13916   47564   81056   97025   85977   29372

The four people assigned are

(A)   Zavala, Liu, Lonnquist, Klein.
(B)   Liu, Harris, Klein, Scott.
(C)   Zavala, Lonnquist, Elifritz, Klein.
(D)   Zavala, Lonnquist, Klein, Scott.
(E)   Zavala, McGarry, Klein, Scott.

#17   A child is 40 inches tall, which places her in the top 10% of all children of similar age. The mean height for such children is 38 inches. Based on this information, what is the standard deviation?

(A)   0.20 inches
(B)   0.31 inches
(C)   0.65 inches
(D)   1.21 inches
(E)   1.56 inches

#18   A study was done to compare the effectiveness of two different groups of teachers. Fifteen different measures were used. In all cases, the students of the teachers from Group A scored higher than the students of the teachers in group B. The report of the study stated that in 2 of the 15 cases the sample size was too small to verify that the difference was statistically significant. This is an illustration of

(A)   good replication.
(B)   placebo effect.
(C)   low power.
(D)   stratified sampling.
(E)   marginal frequency.

#19   Traffic engineers studied the traffic patterns of two adjacent busy intersections at rush hour. At the first intersection, the average number of cars waiting to turn left was 17 with a standard deviation of 4 cars. At the second intersection, there are 25 cars waiting to make a left turn with a standard deviation of 7 cars. The report combined the mean number of cars at both intersections. Assuming the data are independent, the standard deviation of the sum is

(A) $\sqrt{11}$ cars.
(B) $\sqrt{33}$ cars.
(C)   8 cars.
(D) $\sqrt{65}$ cars.
(E)   11 cars.

#20   A newspaper wants to know public opinion of a town regarding the construction of a new library in a downtown location. It is decided that 48 people will be surveyed using a simple random sample. Which of the following will produce a simple random sample?

(A)   Record the opinions of the first 48 people who visit the newspaper's web site.
(B)   Survey every fourth person who enters the current library until 48 people have responded.
(C)   Randomly select 48 people from the city phone directory.
(D)   Randomly select 12 people from each of the northwest, northeast, southwest, and southeast sections of the city.
(E)   Number the residents using the latest census data. Use a random number generator to pick 48 people.

#21   Suppose the probability that a softball player gets a hit in any single at-bat is .300. Assuming that her chance of getting a hit is independent, what is the probability that she will not get a hit until her fourth at-bat in a game?

(A)   0.0081
(B)   0.1029
(C)   0.3430
(D)   0.4116
(E)   0.7599

#22   Which of the following statements is FALSE?

(A)   The power of a hypothesis test increases as α increases.
(B)   The power of a hypothesis test does not depend on the sample size.
(C) β is a measure of a Type II error.
(D)   The power of a hypothesis test is a measure of the ability of the test to detect a difference between the estimated value and the true value of a parameter.
(E)   As the α-level increases, the β-level of a hypothesis test decreases.

typing speeds are approximately normally distributed with a mean of 55 words per minute with a standard deviation of 4 words per minute. If Kim scores at the 95th percentile, then his typing speed is closest to

(A)   47 wpm.
(B)   50 wpm.
(C)   59 wpm.
(D)   62 wpm.
(E)   67 wpm.

#24   A new dietary supplement was given to a volunteer group of 25 people who were at least 50 pounds overweight. Eighteen people from the group lost at least 20 pounds. What can we conclude about the effectiveness of the diet?

(A)   The diet was effective in 72% of the cases.
(B)   The mean weight loss was more than 20 pounds.
(C)   Nothing, since the 25 people were not an SRS.
(D)   Nothing, since there was no control group that received a placebo.
(E)   Nothing, since the people knew they were using the diet.

#25   A pharmaceutical company develops a new medication to lower a person's cholesterol. The advertisement for the new drug cites the results of an experiment comparing the drug with a placebo stating the level had been lowered by an average of 13.2 mg/l with a   P-value < 0.01. Which of the following best explains the meaning of P-value?

(A)   1% of the people did better with the placebo than with the drug.
(B)   Only 1% of the people experienced side effects.
(C)   The probability that a group of people randomly selected from the control group would have a mean decrease in cholesterol of at least 13.2 mg/l is less that 1%.
(D)   All but 1% of the people on the drug experienced a drop of 13.2 mg/l.
(E)   The difference between the two groups was not significant.

#26   A linear regression was conducted on the following data points: A(1,11), B(3, 7), C(5, 2), D(7, 1), and E(9, -2). The residual for which point has the largest absolute value?

(A)   A
(B)   B
(C)   C
(D)   D
(E)   E

11

#27   An alumni association of a major university finds that 30% of its members attained advanced degrees, 50% attained a bachelor's degree and 20% did not graduate. The mean income of a graduate with an advanced degree is $75,000, the mean income of a graduate with a bachelor's degree is $54,000, and the mean income of a non-graduate is $36,000. From past results, alumni members donate 5% of their income to the university. If 50 members of the alumni association are randomly selected, what is the expected total donation?

(A) $2,750
(B) $2,835
(C) $137,500
(D) $135,000
(E) $141,750

#28   In order to determine the dexterity or handedness of people, 10 people were asked to roll a die with one hand and make a tally mark with the other hand for 60 seconds. Then the subjects were asked to use the other hand. The hand that each person used first was decided by chance. The number of tally marks made by each subject is shown in the table below.

Subject	1	2	3	4	5	6	7	8	9	10
Right hand	42	36	50	42	32	41	40	37	38	47
Left hand	45	49	31	41	44	53	36	48	52	33

One measure of handedness is the difference between the number of marks made by each hand. What is the P-value for the appropriate hypothesis test?

(A)   -0.910
(B)   0.376
(C)   0.493
(D)   0.715
(E)   0.754

#29   A researcher made a scatter plot from some previously collected data. The data was clearly non-linear in shape. The researcher then tried a variety of transformations on the data in an attempt to linearize the results. The residual plot for each is shown below.

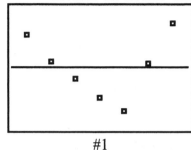

#1

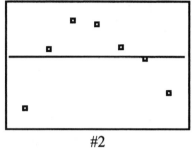

#2

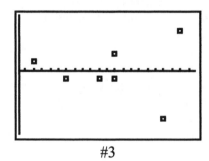

#3

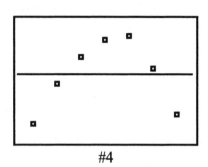

#4

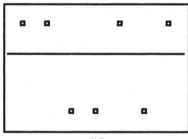

#5

Which of the transformations was best at linearizing the data?

(A)   #1
(B)   #2
(C)   #3
(D)   #4
(E)   #5

#30   A spring is used to propel a toy rocket. In a sample of ten toy rockets, the mean distance traveled is 32.7 inches with a standard deviation of 4.3 inches. What is a 98% confidence interval for the true mean of the distance traveled?

(A) 32.7 ± 3.16
(B) 32.7 ± 3.21
(C) 32.7 ± 3.26
(D) 32.7 ± 3.76
(E) 32.7 ± 3.84

#31   A student is assigned a project in a statistics class. The student uses the Internet to find the data from the 2000 Census that lists the population for each state in the United States and computes the 90% confidence interval for the mean population of each state. The student's statistics teacher marks the answer wrong. Which of the following is the reason the student's answer is incorrect?

(A)   The data does not represent a true random sample.
(B)   Since the data are highly skewed, it violates the rules of thumb to construct a confidence interval.
(C)   A confidence interval cannot be constructed when population data is known.
(D)   The degrees of freedom is unknown.
(E)   Using mean from each state to compute the overall mean is improper.

#32   A survey of 259 families was made to determine their vacation habits. The two-way table below shows the number of families by location (rural, suburban, urban) and length of most recent vacation (1 - 7 days, 8 or more).

	Rural	Suburban	Urban	Total
1-7 days	80	39	37	156
8 or more days	52	32	19	103
Total	132	71	56	259

What is the probability that a family selected was suburban, given that they spent eight or more days on vacation?

(A)   0.12
(B)   0.31
(C)   0.45
(D)   0.50
(E)   28.24

#33   A large bakery has many different products for sale. Suppose that 70% of all customers of the bakery order donuts, 50% order cinnamon rolls, and 40% order both. If a customer is randomly selected, what is the probability that she ordered neither donuts nor cinnamon rolls?

(A)   14%
(B)   20%
(C)   35%
(D)   48%
(E)   60%

#34   A group of eight people work on the fifth floor of a building. They all decide they want to lose weight by increasing their physical activity. Four of them decide to walk up the five flights of stairs each work day. The other four decide to walk down the five flights at the end of the day. They weigh themselves each Monday morning and report their weight to their supervisor. This is an example of

(A)   an experiment since the workers are divided into two equal groups.
(B)   an observational study since the supervisor didn't decide which group walked up the stairs.
(C)   block design since there are two treatments.
(D)   a survey since the people volunteered to report their weight each week.
(E)   an explanatory variable since both groups increased their physical activity.

#35   The histogram below represents the yield (in bushels/acre) of 20 apple orchards in the northwest. The class intervals are equal. The right endpoint is included but the left endpoint is not.

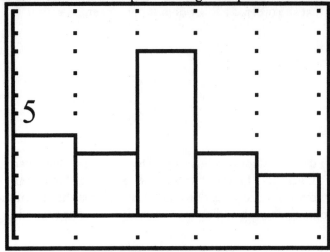

Bushels/acre (0-100)

What proportion of orchards exceeded 60 bushels/acre?

(A)   3%
(B)   5%
(C)   15%
(D)   25%
(E)   60%

#36   An advertisement for a diet supplement, Looze-Now, claims that it is scientifically proven to help people lose an average of 14.8 pounds in 30 days.   The article further explains that a double blind test was done to compare Looze-Now with a control group. The mean weight loss of people on Looze-Now was 14.8 pounds with a standard deviation of 11.2 pounds. The mean weight loss of people in the control group was 15.1 pounds with a standard deviation 12.5 pounds. Which of the following statements is true?

(A)   Looze-Now caused people to lose weight.
(B)   The least amount of weight lost among the Looze-Now group was 3.6 pounds.
(C)   The weight loss among the Looze-Now sample was likely due to a placebo effect.
(D)   The experiment wasn't really double blind since the people could tell if they were losing weight.
(E)   Anyone who takes Looze-Now will experience a weight loss.

16

#37 Automobile engineers checked the effect of a new automatic transmission on mileage. They selected 50 cars of the same make and model. They randomly chose 25 of the cars to use the old transmission, while the other 25 cars received the new model of transmission. The mean of the first group was 23.4 mpg with $s = 3.4$ mpg. The mean of the second group was 25.1 mpg with $s = 3.2$ mpg. A hypothesis test on the difference between the means was conducted. Which of the following is an assumption that allowed for the hypothesis test?

(A) The choice of transmission was random.
(B) The sampling distribution of the difference between the sample means is bimodal.
(C) Since the data was random, no linear regression was possible.
(D) The standard deviation of each group was different.
(E) The difference between the means was within a standard deviation.

#38 A study was done at a large mid-western university to determine if there was bias in the starting salaries of short people when compared to people of above average height. Since the mean height for women is different from the mean height for men, the researchers examined both men and women. Graduates were selected from five different majors. For each major, a randomly selected woman at least 2 inches below the mean height for women was paired with a randomly selected woman who was at least 2 inches above the mean height. Five pairs were similarly selected from male graduates from the same majors. The starting weekly salaries for each person in the pair was recorded and the difference (Tall Salary - Short Salary) computed. The results were summarized in the computer output below.

	N	MEAN	STDEV	SE MEAN	95 PERCENT C.I.
DIFF	10	420.000	473.286	149.666	(83.431, 758.574)

Based on the data, which of the following statements is correct?

(A) A chi-squared test should have been done since there were two factors -- height and gender.
(B) Since the confidence interval did not contain 0, it is likely that the mean salary for tall people is greater than the mean salary for short people.
(C) There is a 95% probability that the true mean difference in salaries lies between $83.43 and $758.57.
(D) Since the confidence interval was above 0, it is likely that the mean salary for tall people is greater than the mean salary for short people.
(E) A sample size of 10 is too small to draw any conclusion.

#39   The following graph represents the number of jobs related to the 2002 Winter Olympics from 1996 through 2003.

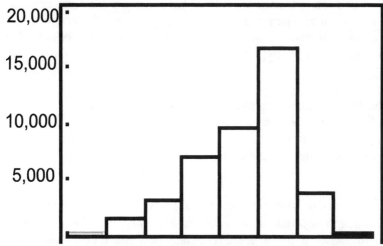

1996 1997 1998 1999 2000 2001 2002 2003

Which of the following statements are TRUE?
I.   The data is skewed to the left.
II.   The data is skewed to the right.
III.   The mean number employed is about 5,000.
IV.   The mean number is about 10,000.

(A)   I and III
(B)   II and III
(C)   I and IV
(D)   II and IV
(E)   None of the above are true.

#40   In a study children were asked to indicate which emotion they associated with the color blue. The response and gender of the child are summarized in the table

	Joy	Happiness	Love	Anger	Total
Male	28	20	40	18	106
Female	61	25	80	60	226
Total	89	45	120	78	332

We wish to test the claim H_0: There is a relationship between gender and emotion vs. H_a: There is no relationship between gender and emotion. (Let $\alpha = 0.05$)

(A)   Reject H_0: If $\chi^2 = 0.103$
(B)   Reject H_0: If $\chi^2 = 6.17$
(C)   Reject H_0: If $\chi^2 = 7.81$
(D)   Fail to reject H_0: If $\chi^2 = 14.07$
(E)   Fail to reject H_0: If $\chi^2 = 15.51$

18

#1

In hilly areas farmers often contour their fields to reduce the erosion due to water flow. This might have the unintended effect of changing the yield since the rows may not be aligned in an east-west direction to take advantage of direct sunlight.

The following summary statistics show the yields in hundreds of bushels per acre of randomly selected plots of land along hillsides as well as level plots of land nearby with similar soil characteristics.

Quantiles

Level	Minimum	10.0%	25.0%	Median	75.0%	90.0%	Maximum
Hilly	10.9	10.9	13.3	29.2	58.1	97.4	100.0
Level	10.9	11.2	18.5	39.3	61.5	103.4	118.0

Means and Standard Deviations

Level	Number	Mean	Std Dev	Std Err Mean
Hilly	12	39.7	30.4	8.8
Level	12	44.6	31.1	9.0

(A)   Draw side-by-side box plots that show the yields of the two types of land.

(B)   Describe the shape and any unusual features of both box plots.

(C)   Based on your plots, what can you conclude about the types of land?

#2

A fast food chain, YummyStuff, wants to introduce a new product. They select a city that has 16 YummyStuff stores to be a test market. The advertisers have developed two different in-store advertising campaigns to promote the new product. One is designed to stress convenience and fun. The other is designed to stress value. From previous studies it is known that the median income of the neighborhoods surrounding the eight stores east of Main Street is 50% higher than the median income from the neighborhoods surrounding the eight stores west of Main Street.

(A)   Describe an experimental design that would test the effectiveness of each campaign for both income levels. Be specific in your explanation. You may use diagrams as part of your explanation.

(B)   Describe any possible confounding variables.

(C)   Explain what type of hypothesis test you would use to determine which campaign was most effective.

The average January minimum temperature (°F) for 15 selected U.S. cities is shown in the table below along with the city's latitude and longitude coordinates in degrees.

City	Avg Temp	Latitude	Longitude	City	Avg Temp	Latitude	Longitude
Albany, NY	14	42.6	73.7	Miami, FL	58	26.3	80.7
Bismarck, ND	0	47.1	101	Mobile, AL	44	31.2	88.5
Boise, ID	22	43.7	117.1	Phoenix, AZ	35	33.6	112.5
Boston, MA	23	42.7	71.4	Portland, OR	33	45.6	123.2
Denver, CO	15	40.7	105.3	Seattle, WA	33	48.1	112.5
Houston, TX	44	30.1	95.9	Salt Lake City, UT	18	41.1	112.3
Los Angeles, CA	47	34.3	118.7	Wilmington, DE	26	40.5	76.3
Milwaukee, WI	13	43.3	88.1				

Meteorologists want to determine if there is a relationship between the temperature and either the city's latitude or longitude.

Scatter plots for both relationships are shown on the next page along with the equation for the least squares regression line.

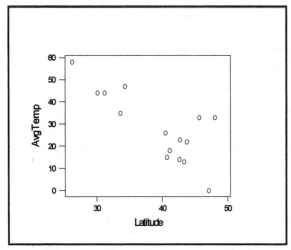

AVGTEMP = 100.5 - 1.83 * LAT
R-SQ = 0.61   R-SQ (ADJ) = 0.65

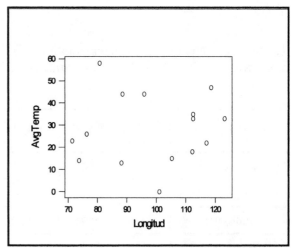

AVGTEMP = 23.5 + 0.0498 * LONG
R-SQ = 0.0031   R-SQ (ADJ) = 0.0053

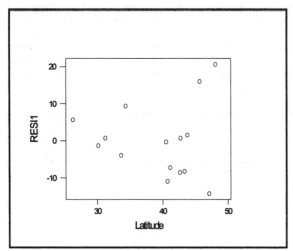

RESIDUAL PLOT
LATITUDE VS. RESIDTEMP

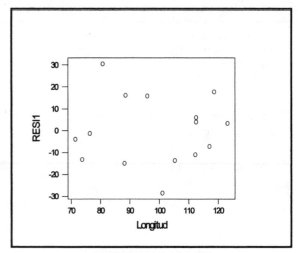

RESIDUAL PLOT
LONGITUD VS. RESIDTEMP

(A)   Which scatter plot shows a stronger relationship between average temperature and map location? Explain your reasoning.

(B)   What is the value of the correlation coefficient for latitude and average temperature?

(C)   Chicago, IL is at 42.3° latitude and 88° longitude. Predict the average January minimum temperature. Show your work.

(D)   What is the residual for longitude and average temperature for Denver, CO? Show your work.

24

(E) A hypothesis test for the slope between latitude and average temperature was performed. The t-statistic of the hypothesis test $H_o: \beta = 0$ vs. $H_a: \beta \neq 0$ is -4.495 with degrees of freedom = 13. Is this strong or weak evidence to reject the null hypotheses? Why?

#4

The administrators of a large hospital installed new drink vending machines in the cafeteria and main entrance in addition to the ones in the waiting room. They want to know if flavor preference depends on location. A random sample of visitors who purchased drinks from vending machines was selected and interviewed. The results are shown in the table below.

	Cafeteria	Main entrance	Waiting area
Cola	33	57	112
Juice	30	20	93
Water	5	35	83

(A)   What is the expected value for sales of juice in the waiting area? Show your work.

(B)   What is the probability that a person buys juice, given that they purchase a drink in the cafeteria?

(C)   Write the null and alternate hypotheses for an analysis of the administrators' question.

(D)   Determine the value of the appropriate test statistic, the degrees of freedom, and the P-value.

(E)   Explain what you would tell the administrators regarding drink preference and location in this situation.

26

#5

Corrinna is a basketball player who makes 80% of her free throws. In an important game, she shoots 12 free throws and makes 8.

(A)   Assuming the success rate of free throws is independent, what is the probability that Corrinna will make exactly 8 of 12 free throws?

(B)   Find an approximate 95% range for the number of free throws that Corrinna will usually make out of 12 attempts.

(C)   Some fans think that Corrinna missed so many shots because the game was important and she was so nervous that it affected her ability. Based on the results from parts (A) and (B), is this conclusion warranted? Explain your reasoning.

(D)   Suppose that over the next 90 shots, Corrinna made 60 of them. Does this information still support the claim that she is successful in 80% of her shots? Why?

#6

For surveying purposes, land in the United States is divided into townships. Each township is a square 6 miles on a side and is divided into 36 squares, each of which is called a section. The sections are numbered according to the following design.

6	5	4	3	2	1
7	8	9	10	11	12
18	17	16	15	14	13
19	20	21	22	23	24
30	29	28	27	26	25
31	32	33	34	35	36

At the time of statehood, the federal government gives states one randomly selected section within each township. All the revenues from that section are to be used to support local schools.

(A)   Explain why this numbering system can be used to create a Simple Random Sample.

(B)   Suppose that schools received $1,000 from each section that was not dedicated for school revenues. Suppose further that schools received $100,000 from the section that is dedicated for school revenues. If a section is selected at random, what is the expected income from that section?

(C)   Let Section #12 be the section given to the state for school support. Using the random number table below, draw two random samples (without replacement) of five sections each and determine the total revenue received by each sample. Explain your process and show clearly the numbers you select.

(D)   Suppose that many simple random samples of size n = 5 were drawn (without replacement) and the total revenue received computed for each sample. If the totals for each sample were plotted, what would be the mean of the distribution?

29676	20591	68086	26432	46901	20849	89768	81536	86645	12659
92259	57102	80428	25280	00742	57392	39064	66432	84673	40027
32832	61362	98947	96067	64760	64584	96096	98253	05366	04213

Practice Exam 2

#1   The box plot of a given set of data is shown here.

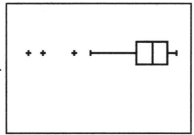

Which of the following statements is TRUE?

(A)   The data are strongly skewed to the right with three outliers.
(B)   The data are uniformly distributed and the outliers do not affect the median.
(C)   The data are normally distributed with three outliers.
(D)   Only two of the outliers are less than {Quartile1 - 1.5* InterQuartile Range}.
(E)   The data are strongly skewed to the left with three outliers.

#2   A radio station is interested in predicting the proportion of registered voters who support an increase in the state sales tax. Listeners were asked to go to the station's web site and indicate whether they favored an increase to support additional regional parks. 1744 listeners logged on and 922 (53%) were against the increase. The population of interest is

(A)   the 1744 listeners who logged on.
(B)   all registered voters.
(C)   all people who listen to the station.
(D)   only that portion of the 1744 listeners who voted that are actually registered to vote.
(E)   the 922 listeners who were against the proposed increase.

#3   A recent newspaper article cited a poll that reported 4% of the adults in Wyoming believed that overcrowding was a serious problem, while 36% of respondents in California had the same opinion. The reporter said that an average of 20% of the population of the two states felt that overcrowding was therefore a problem. A statistician criticized the article stating the conclusion was inaccurate. Which of the following reasons support the statistician's reasoning?

(A)   The number of people in the two states are very different.
(B)   The polls did not represent an SRS from each state. Wyoming has less overcrowding than California, so the results are biased on this issue.
(C)   The correct value should have been 40%.
(D)   The sample size from each state was not large enough to support a 95% confidence interval.
(E)   The standard deviations of the two states are different.

#4   One of the topics studied by The Marriage Project at Rutgers University was "I am confident that I can find a marriage partner that will meet my needs in a marriage." Preliminary research showed that 88% of respondents agreed with the statement. How many people must be sampled to estimate the true proportion within ±0.04 with 95% confidence?

(A)   178
(B)   254
(C)   391
(D)   601
(E)   1201

#5   Sixty-eight people from a random sample of 128 residents of Uppsala, Sweden, had blue eyes. Forty-five people from a random sample of 110 people from Preston, England, had blue eyes. Let p_1 represent the proportion of people in Uppsala with blue eyes and p_2 represent the proportion of people in Preston with blue eyes. What is the P-value of the hypothesis test: H_0: $p_1 = p_2$ vs. H_a: $p_1 \neq p_2$?

(A)   0.03
(B)   0.06
(C)   0.41
(D)   0.47
(E)   0.53

#6   A random sample of 100 homes counted the number of television sets in each home. The results are shown in the histogram below.

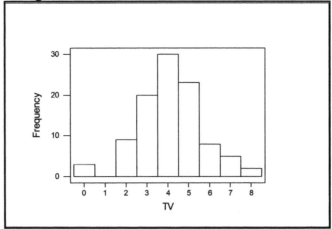

Which of the following normal probability plots also represents this same data set?

(A)

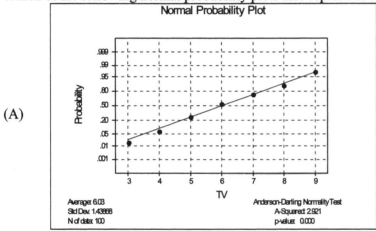

(B)

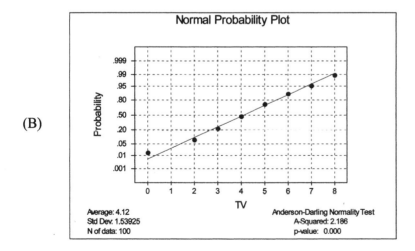

(Continued on next page)

(C)

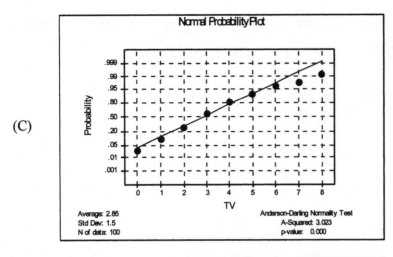

(D)

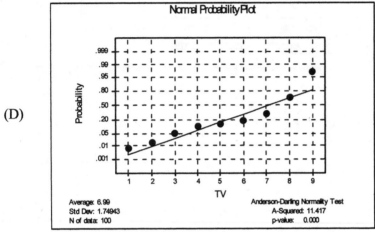

(E)

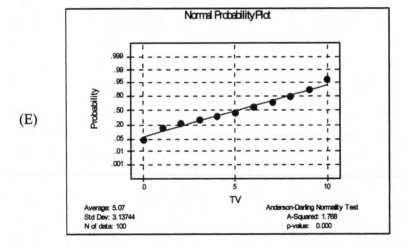

#7   An efficiency expert wanted to see if there is a relationship between the number people invited to a meeting and the number of minutes late that the meeting started. The table shows the results with the accompanying scatter plot.

Number of people invited	1	2	3	4	5	10
Number of minutes late the meeting started	3	6	8	10	14	29

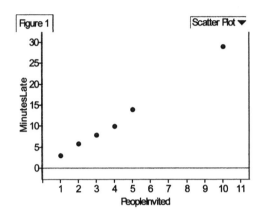
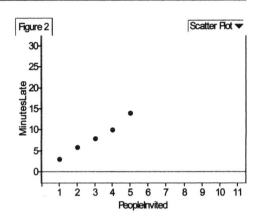

Figure 2 represents the scatter plot with the point (10, 29) removed from the data. Which of the following is NOT true about the point (10, 29).

(A)   It is an influential observation.
(B)   It is has a small residual.
(C)   It is an outlier.
(D)   The P-value of a Linear Regression t-test changes little when the point is removed.
(E)   None of the above.

#8   A sports physiologist wants to compare the effects of two different exercise machines on the flexibility of gymnasts. There are 20 gymnasts available. She decides to group the gymnasts in matched pairs. Which of the following would give the least biased result?

(A)   Use a random number table to divide the gymnasts into ten separate pairs. Then assign the first person to the first machine and the second person to the second machine.
(B)   Use a random number table to divide the gymnasts into ten separate pairs. Then flip a coin to assign each member of a pair gymnasts to the two machines.
(C)   Use a random number table to divide the gymnasts into two groups. Flip a coin to assign each group to the machines.
(D)   Pair the gymnasts subjectively (keeping both members of the pair as similar as possible). Use a random number table to assign each member of the pair to a machine.
(E)   Use a random number able to divide the gymnasts into pairs. Have both gymnasts use both machines, using a coin flip to decide which machine is used first.

#9   The number of goals scored per game in a full season of soccer games for a professional league is strongly skewed to the right with a mean of 2.3. An SRS of sample size n = 30 is taken from the population and the sample mean is computed. This is repeated for a total of 175 trials. Which of the following best describes the shape of the sampling distribution?

(A)   Skewed to the right with a mean of 2.3 goals

(B)   Skewed to the right with a mean of $\dfrac{2.3}{\sqrt{30}}$ goals

(C)   Binomially distributed with a mean of 2.3 goals

(D)   Normally distributed with a mean of $\dfrac{2.3}{\sqrt{30}}$ goals

(E)   Normally distributed with a mean of 2.3 goals

#10   Students in a statistics class drew circles of varying diameters and counted how many Cheerios® could be placed in the circle. The scatter plot shows the results.

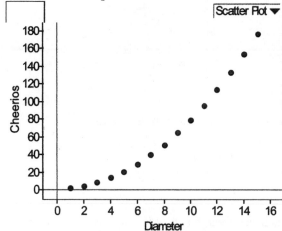

The students wanted to determine an appropriate equation for the relationship between diameter and the number of Cheerios®. The students decided to transform the data to make it appear more linear before computing a least squares regression line. Which of the following transformations would be reasonable for them to try?

I.   Take the square root of the number of Cheerios®.
II.   Cube the number of Cheerios®.
III. Take the log of the number of Cheerios®.
IV.   Take the log of the diameter.

(A)   I and II
(B)   I and III
(C)   II and III
(D)   II and IV
(E)   III and IV

#11   To determine the viscosity of a liquid, scientists measure the length of time it takes for a fixed amount of the liquid to run through a tube. Two different types of motor oil were tested with 30 trials each. The first type of oil had a mean time of 10 seconds with a standard deviation of 2 seconds. The second type of oil had a mean of 14 seconds with a standard deviation of 3 seconds. The difference between the mean time of the two types of oil is 4 seconds with a standard deviation of

(A) $\frac{1}{4}$ seconds.

(B) $\frac{5}{\sqrt{20}}$ seconds.

(C) $\sqrt{13}$ seconds.

(D)   5 seconds.

(E) $\frac{5}{\sqrt{40}}$ seconds.

#12   According to data from Major League Baseball, the mean salary for professional major league baseball players has been increasing for the past 15 years. A sportswriter looks at the data for all players for the 2002 season and decides to construct a 95% confidence interval for the mean salary for all players. Which of the following is the reason the writer's reasoning is incorrect?

(A)   The sportswriter did not take a Simple Random Sample.
(B)   A confidence interval cannot be constructed when population data is known.
(C)   Since a very few ball players make over $20,000,000 a year, the data is highly skewed, which violates the rules of thumb to construct a confidence interval.
(D)   Since the number of players is not listed, the degrees of freedom is unknown.
(E)   Using the data from only one year is not sufficient to verify the trend from previous years.

#13   A random variable X has a probability distribution as shown in the graph. What is the probability that X will be greater than or equal to 1.75?

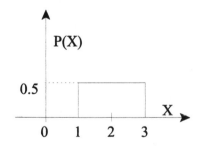

(A)   0.250
(B)   0.375
(C)   0.625
(D)   0.750
(E)   1.750

#14   Two different brands of blood pressure cuffs were tested on the same population. The results are shown in the table below.

	Brand A	Brand B
Sample mean	127	134
Standard deviation	23	20
Sample size	20	17

What is the 95% confidence interval for the difference between the two means?

(A) -7 ± 3.23
(B) -7 ± 11.63
(C) -7 ± 12.34
(D) -7 ± 13.85
(E) -7 ± 14.98

#15   At a fund raiser for charity 15 college basketball players were matched with 15 people who were not basketball players who gave a donation to the charity. Each person then shot ten free throws and the difference between the player and the partner was recorded (College player - Partner). The results are summarized in the computer output below.

	N	MEAN	STDEV	SE MEAN	95 PERCENT C.I.
DIFF	15	2.600	3.290	0.849	(0.77768, 4.42232)

Based on the information, which of the following statements is TRUE?

(A)   There is a 95% probability that the true mean difference between college players and the randomly selected partners is between 0.78 and 4.42.
(B)   Since the confidence interval did not contain 0, it is likely that the mean number of shots made by college players is greater than the mean number of shots made by randomly selected people.
(C)   A sample size of 15 is too small to draw any conclusion.
(D)   Since there is no relationship between the two groups, a two sample t-test is more appropriate than a one sample t-test on the difference.
(E)   All college players made more shots than their partner.

#16   Suppose that 40% of the cars in a certain town are white. A person stands at an intersection waiting for a white car. Let X = the number of cars that must drive by until a white car drives by. The probability that a white car drives by before $X = 5$ is

(A)   0.0518.
(B)   0.1296.
(C)   0.2592.
(D)   0.8704.
(E)   0.9482.

#17   According to the Central Florida Blood Bank, 13% of first-time donors return to make a second donation within three months. In the fall of 2001 they tracked 6000 first-time donors and found that 891 donated a second time within 90 days. What is the probability that the rate of repeat donations among the sample group was greater than the long-term rate?

(A)   Less than 0.001
(B)   0.007
(C)   0.014
(D)   0.492
(E)   0.983

#18   The mean blood pressure for 47 year-old males in the United States is 139 mg with a standard deviation of 26 mg. A doctor tells his 47 year-old male patient he is in the lowest 10% of all people in this population. Which of the following is nearest to the patient's blood pressure?

(A)   96
(B)   106
(C)   108
(D)   125
(E)   127

#19   A pharmacist wants to study the effect of temperature (0° F, 30° F, 60° F, and 90° F) on the potency of a headache pain reliever when it is stored for extended periods of time. Thirty pills were randomly assigned to each of the temperatures for the specified storage time. The time it took for a person to gain relief from a headache was measured for each pill. Which of the following is the correct description of the treatment, experimental unit, and response?

(A)   Specific temperature, relief time, pill
(B)   Specific temperature, the pharmacist, relief time
(C)   Pill, relief time, specific temperature
(D)   Random assignment, pill, relief time
(E)   Specific temperature, pill, relief time

#20   A statistics class took a random sample of the students at the school to find the proportion of those who claimed to be vegetarians. Another statistics class in another school took a similar random sample of the students at its school. The results from the first school were $n_1 = 75, \bar{x}_1 = 12$. The results from the second school were $n_2 = 45, \bar{x}_2 = 9$. What is the 90% confidence interval for the difference between the population proportions of the two schools that claim to be vegetarians?

(A) -0.4 ± 0.009
(B) -0.4 ± 0.01
(C) -0.4 ± 0.06
(D) -0.4 ± 0.12
(E) -0.4 ± 0.14

#21   A company with 16 employees gives everyone a $2000 bonus. What will be the change in the standard deviation of the employees' income after the bonus is awarded?
      (A)   It will increase by $2000.
      (B)   It will stay the same.
      (C)   It will increase by $500.
      (D)   It will increase by $\sqrt{2000}$.
      (E)   It will be multiplied by $2000.

#22   A local dealer has two video stores in a town, one on Foothill Drive and the other on Grand Avenue. The Foothill Drive store does 70% of the dealer's business in the town, and the Grand Avenue store does the rest. In the Foothill Drive store 40% of all rentals are DVDs. At the Grand Avenue store 30% of all rentals are DVDs. If a customer is selected at random, what is the probability that she rented a DVD?

      (A)   0.175
      (B)   0.33
      (C)   0.35
      (D)   0.37
      (E)   0.70

#23   In order to conserve water in a dry state, residents are asked to measure how much water their sprinkler system discharges in 15 minutes. They are then asked to water according to the following schedule.

Water depth in inches in 15 minutes	0.125	0.187	0.25	0.313	0.375	0.50	0.625	Total
Total weekly watering time (in minutes)								
Spring	52	34	26	20	17	13	10	172
Summer	104	69	52	41	35	26	21	348
Fall	69	51	39	31	26	19	15	250
Total	225	154	117	92	78	58	46	770

How much does the cell corresponding to 0.25 inches of water in 15 minutes with Summer watering contribute to the total chi-square statistic?

      (A)   -0.017
      (B)   0.015
      (C)   0.515
      (D)   0.999
      (E)   52.88

#24   The following scatter plot shows the mean skid length for a car weighing 1500 pounds traveling at 40 mph on a new set of tires when plotted against the outside temperature (in C°). Which of the following statements is TRUE?

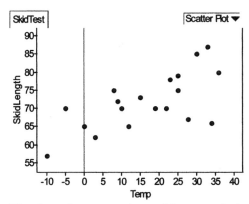

(A)   The data show a weak, positive correlation.
(B)   The data show a strong, positive correlation.
(C)   The data show a weak, negative correlation.
(D)   The data show a strong, negative correlation.
(E)   No linear relationship exists.

#25   A sales person ranks in the top 5% of all sales people in a large company. If the annual mean sales amount is $750,000 and the standard deviation is $150,000, how much does the person sell each year?

(A) $757,500
(B) $996,750
(C) $1,044,000
(D) $1,697,100
(E) $2,650,500

41

#26   The following information is a summary about a data set from pulse rates for 3 year-old children.

```
1-Var Stats
 x̄=97.00
 Σx=2910
 Σx²=290518
 Sx=11.93931
 σx=11.73864
 n=30
 MinX=74
 Q1 = 88
 Med=98.5
 Q3=107
 MaxX=118
```

What is the smallest integer value for an outlier on the upper end?

(A)   120
(B)   125
(C)   127
(D)   136
(E)   173

#27   Twenty-five hamsters were trained to run a maze. The times for each hamster to complete the maze were normally distributed with a mean of 48 seconds and a standard deviation of 10 seconds. What is the range for the middle 95% of all times to run the maze?

(A)   18 to 78 seconds
(B)   28 to 68 seconds
(C)   38 to 58 seconds
(D)   44 to 52 seconds
(E)   46 to 50 seconds

#28   A random sample of cell phone customers for a company was taken from each of two large cities to measure cell phone usage. The number of calls in a month for each customer was recorded. The results are summarized below.

Quantiles

Level	Minimum	10%	25%	Median	75%	90%	Maximum
City A	1.9	1.9	4.3	20.2	58.1	97.4	100.0
City B	1.9	2.2	9.5	20.3	61.5	103.4	141.0

Means and standard deviations

Level	n	Mean	Std Dev	Std Err Mean
City A	12	39.7	30.4	8.8
City B	12	44.6	31.1	9.0

Which of the following statement are correct?
I.   Customers in City A, on average, make fewer cell phone calls than those in City B.
II.   The data from City A are more strongly skewed than the data from City B.
III.   The maximum from City B is an outlier.

(A)   I only
(B)   II only
(C)   III only
(D)   I and II
(E)   I and III

#29   A company has 50 sales people. A survey of ten representatives found the mean years with the company was 8.7 years with s = 3.2 years. If the survey is an SRS of the company's sales force, what is the 95% confidence interval for the mean years of service of all sales people?

(A) 8.7 ± 1.98
(B) 8.7 ± 2.09
(C) 8.7 ± 2.29
(D) 8.7 ± 2.65
(E) 8.7 ± 2.79

#30 A researcher wishes to test the yield of a new type of fertilizer. He divides a small field into 100 plots as shown in the diagram. Each row gets a different concentration of fertilizer. The design of the experiment requires that ten different plots be sampled.

1	2	3	4	5	6	7	8	9	10
11	12	13	14	15	16	17	18	19	20
21	22	23	24	25	26	27	28	29	30
31	32	33	34	35	36	37	38	39	40
41	42	43	44	45	46	47	48	49	50
51	52	53	54	55	56	57	58	59	60
61	62	63	64	65	66	67	68	69	70
71	72	73	74	75	76	77	78	79	80
81	82	83	84	85	86	87	88	89	90
91	92	93	94	95	96	97	98	99	100

East

Because there is a strong wind that blows from the east, the researcher decides to randomly select one plot in each row (1-10, 11-20, 21-30, etc). This type of sampling is known as

(A)   multi-stage sample.
(B)   simple random sample.
(C)   survey sample.
(D)   convenience sample.
(E)   stratified sample.

44

#31   The resting pulse rate of 11 people was measured before and after a televison commercial was shown them. The researcher wanted to check the mean difference in pulse rate for each subject. The results are shown in the following table.

Subject	1	2	3	4	5	6	7	8	9	10	11
Pulse rate before	85	79	73	76	75	79	74	67	80	69	71
Pulse rate after	83	78	85	82	81	90	75	69	73	82	70

What is the degrees of freedom for the appropriate t-test for testing if the mean increase is greater than zero?

(A)   10
(B)   11
(C)   20
(D)   21
(E)   22

#32   A researcher wanted to measure the sensitivity of a hypothesis test to detect difference between the mean of a sample and the known mean of the item. When the difference was great, she could use a small sample size. When the difference between the means was small, she had to increase the sample size. This is an illustration of
(A)   a two-sample t-test.
(B)   power.
(C)   a correlation coefficient.
(D)   a type II error.
(E)   convenience sampling.

#33 A strength coach wants to determine if there is a relationship between the weight of an individual and how much weight the individual can lift. The data was collected and analyzed using a statistics software program. The output is shown below.

TotalWeightLifted	StudentWeight		Total
	Under 150	Over 150	
Under 200	44 40.0 0.40	43 47.0 0.34	87 87.0 0.00
Over 200	25 29.0 0.55	38 34.0 0.47	63 63.0 0.00
Total	69 69.0	81 81.0	150 150.0

The number of rows with at least one missing value is 0

Chi-Square Statistics Section

Chi-square	1.745200
Degrees of Freedom	1.0000
Probability level	0.18641
Phi	.107864
Cramer's V	.107864
Kappa's t-value	1.32106
McNemar's Probability Level	.029049

What is the expected value for the number of students who weigh more than 150 pounds and could lift more than 200 pounds?

(A)   0.47
(B)   34
(C)   38
(D)   63
(E)   81

#34   A scatter plot and least squares regression line are shown in the figure below. If the highlighted
      point (20, 21) is removed from the data set, which of the following statements is TRUE?

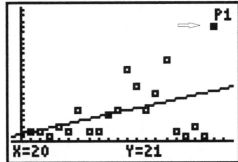

(A)   The slope will decrease and the y-intercept will decrease.
(B)   The slope will increase and the y-intercept will decrease.
(C)   The slope will decrease and the y-intercept will increase.
(D)   The slope will increase and the y-intercept will increase.
(E)   No conclusion can be drawn since the coordinates of the other data points are unknown.

#35   A large bakery has many different products for sale.   Suppose that 70% of all customers of the
      bakery order donuts, 50% order cinnamon rolls, and 40% order both. If a customer is randomly
      selected, what is the probability that he ordered either donuts or cinnamon rolls but not both?

(A)   20%
(B)   24%
(C)   40%
(D)   48%
(E)   60%

#36   A building engineer is trying to determine the wait times for a bank of elevators. He takes a random sample of waiting times. The summary results and box plot of the data are shown below.

$\bar{x} = 1.31$ minutes
$s = 0.85$ minutes
$n = 21$

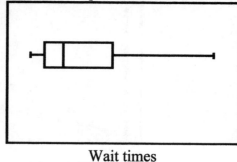

Wait times

Which of the following conditions necessary for an inference test has not been met?
I.   The sample size must be at least 30.
II.   The α level must be stated.
III.   The distribution of the sample data must be approximately normal.

(A)   I only
(B)   II only
(C)   III only
(D)   I and II
(E)   II and III

#37 Two lists of numbers were created. List 1 was random integers from 0 to 25. List 2 was random integers from 5 to 30. The difference of List 1 minus List 2 was sampled 100 times and the differences plotted. Which of the following graphs represents the data?

(A)

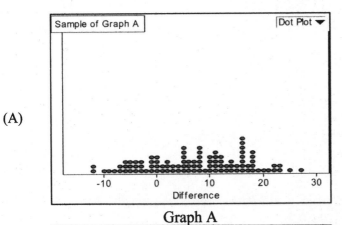

Graph A

(B)

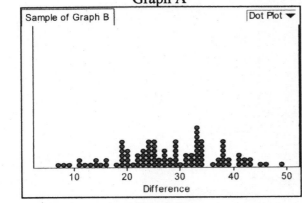

Graph B

(C)

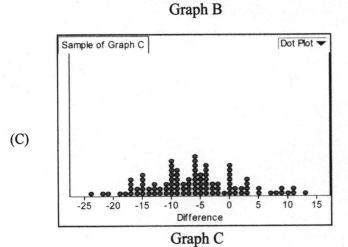

Graph C

(D)

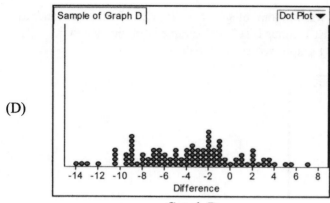

Graph D

(E)

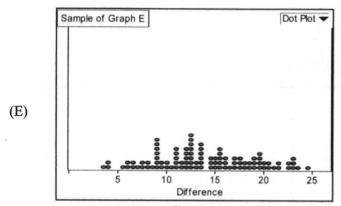

Graph E

#38   An advertiser compared the number of television commercials for a product viewed in one month with viewers' rating of that product. The results are summarized in the table below.

Number of exposures	0-5	6-10	11-15	15-20	21+
Excellent	23	29	42	48	51
Good	25	33	44	53	49
Fair	31	29	25	16	10
Poor	38	32	25	18	12

The advertiser decided to use a χ^2-test to see if there was a relationship. The degrees of freedom for this test is

(A)   3.
(B)   4.
(C)   12.
(D)   18.
(E)   19.

#39   A marine biologist wants to determine the best treatment to eliminate a water-borne parasite on salmon. There are 40 Pacific salmon and 60 Atlantic salmon. The biologist wants to check on three different treatments against a control group. A block design is chosen with 10 Pacific salmon and 15 Atlantic salmon in each group. Which of the following is a valid reason to use a block design for this experiment?

(A)   The conclusions can apply equally to both species.
(B)   There is a genetic difference between the two species.
(C)   It is a requirement of a double blind experiment.
(D)   Each species may respond differently to the treatments.
(E)   There might be different parasites in each ocean.

#40   Twenty-three of 65 people randomly selected in an airport report owning a cell phone. Eighteen of 52 people randomly selected at a rock concert reported owning a cell phone. A researcher wants to test the claim that the proportion of cell phone ownership is not the same between the two groups. Assuming the two groups are independent, what is the P-value of this test?

(A)   0.045
(B)   0.090
(C)   0.465
(D)   0.910
(E)   0.931

#1

You are responsible for planning the layout of a new parking lot for a company with 200 employees. Employees are urged to car pool. From the local chamber of commerce you learn the "average number of workers in daily car pool is 2.3."

(A)   Explain why "2.3" cannot be the median or the mode for the variable "average number of workers in a daily car pool."

(B)   For a variety of reasons, only 91% of the workers actually come to work each day. How many parking spaces are needed for employee parking?

(C)   Suppose the mean commute for each car pool is 19.6 miles with a standard deviation of 6.3 miles. If the distribution of miles commuted is normal, what is the probability that two car pools chosen at random each had a commute greater than 22 miles?

#2

Homing pigeons, when released many miles away from their home coop, have the ability to find their way back home. Scientists speculate that pigeons use the sun as a means of navigating back to their coop. But there are some experts who think pigeons use some other way of determining their way back home, citing individual stories of pigeons who found their way home on sunless days.

Suppose that you had 60 pigeons of the same age that had all been raised in the same coop but had never been released from beyond visual distance of the coop. You may assume that all pigeons are healthy and are physically able to fly the required distance.

(A)   Explain the design of an experiment to determine if pigeons use some other method of navigating besides the sun. Be specific in your explanation. You may use diagrams if they are helpful.

(B)   Describe the hypothesis test you would use to test your results from the experiment. List the required assumptions for this test.

#3

An experiment was done to see how closely people could estimate the number of hours of sleep they got each night. Fifteen volunteers were observed falling asleep in a private room in a lab. The actual number of hours of sleep they received was noted. The next morning each participant was asked how many hours of sleep he or she had gotten. The data was plotted and a regression analysis was performed. Results are shown below.

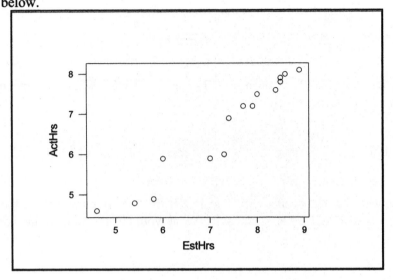

ACTHRS = 0.032 + 0.907 ESTHRS

PREDICTOR	COEF	STDEV	T-RATIO	P
CONSTANT	0.0321	0.4831	0.07	0.948
ESTHRS	0.90744	0.06491	13.98	0.000

S = 0.3198   R-SQ = 93.8%   R-SQ(ADJ) = 93.3%

ANALYSIS OF VARIANCE

SOURCE	DF	SS	MS	F	P
REGRESSION	1	19.988	19.988	195.46	0.000
ERROR	13	1.329	0.102		
TOTAL	14	21.317			

UNUSUAL OBSERVATIONS

OBS	ESTHRS	ACTHRS	FIT	STDEV FIT	RESIDUAL	ST.RESID
6	7.30	6.0000	6.6564	0.0826	-0.6564	-2.12R

R denotes an observation with a large standard residual.

Based on the information, answer the questions on the following page.

(A)   How well were the volunteers able to estimate the actual number of hours of sleep they received. Explain.

(B)   The coefficient of determination is 0.938. Interpret this in the context of this problem.

(C)   Suppose that a person estimated that he had received 8.2 hours sleep when in fact he had actually received 8.5 hours. Compute the residual for this point.

(D)   If the mean number of hours of sleep actually received is 6.70, what is the mean number of estimated hours of sleep?

(E)   Comment on the shape and spread of what the residual plot must look like.

(F)   Suppose the variables had been switched so that Estimated Hours was along the vertical axis and Actual Hours was along the horizontal axis. What would be the value of the correlation coefficient?

#4

A company hires workers to paint ceramic figurines. A new method was proposed to increase the number of figurines painted.

The table below gives the number of figurines painted in a day by a simple random sample of workers before and after the new method was instituted.

Subject #	Figurines painted each day using old method	Figurines painted each day using new method	Subject #	Figurines painted each day using old method	Figurines painted each day using new method
1	30	29	11	20	32
2	28	30	12	29	28
3	31	32	13	31	34
4	26	30	14	29	32
5	20	16	15	34	32
6	30	25	16	20	27
7	34	31	17	26	28
8	15	18	18	25	29
9	28	33	19	31	32
10	20	25	20	29	32

(A)   What is the 95% confidence interval for the mean increase in the number of figurines painted in a day by each worker? Show your computations.

(B)   Based on the 95% confidence interval found in part (A), is a mean increase of 3 figurines per day by each worker a significant improvement? Explain your reasoning.

(C)   Management claims that the new method has increased the production by each worker. Conduct an hypothesis test to determine if the claim is valid.

#5

In order to get truthful responses to potentially embarrassing questions, researchers often ask the question in conjunction with some other activity. Suppose that a large number of adults are interviewed to determine the proportion that have cheated on their income tax. Each person is given a die and placed in a private room. The instructions are as follows: "Roll a die. If you rolled either a 3 or a 4 OR you have ever cheated on your income tax, answer BLUE. Otherwise answer YELLOW."

(A)   Suppose that 10% of the people answer BLUE. What is the approximate proportion of the adults in this survey that have cheated on their income tax?

(B)   Suppose that the results from part (A) came from a random sample of 45 adults. What is the 90% confidence interval for the true proportion of adults who have cheated on their income tax?

#6

There are four different blood types, A, B, AB, and O. The estimated percentage of each blood type in the general population is shown below.

Type	A	B	AB	O
	40%	10%	5%	45%

In addition, 85% of all people are Rh positive which means they carry the Rh factor in their blood. The other 15% are Rh negative, which means that they do not have the Rh factor in their blood.

(A)   Using the random number table below, simulate the selection of 20 people and determine which blood type they have. Also simulate the presence or absence of the Rh factor. Explain your procedure.

Row 1	10480	15011	01536	02011	81647	91646	69179	14194	62590	36207
Row 2	20969	99570	91291	90700	22368	46753	25595	8533	30995	89198
Row 3	27982	53402	93965	34095	52666	19174	39615	99505	24130	48360
Row 4	22527	97265	76393	64809	15179	24830	49340	32081	30680	19655

(B)   Suppose that a simulation found that 9 of 20 people had Type A blood. Furthermore, suppose that the proportion of blood types shown in the table was not known. How many people would need to be sampled to estimate the true proportion of people with Type A blood within 3% at the 96% confidence level?

(C)   Using the information in part (B), what is the 96% Confidence Interval for the estimate of the proportion of people who have Type A blood?

(D)   People with Type O blood can donate blood for transfusion to people with any type of blood group with the restriction that people that are Rh Positive cannot donate to people who are Rh Negative. People who are Rh Negative can donate to both Rh Positive and Rh Negative people. Suppose a blood donor is selected at random. What is the probability that a person with A Negative blood can receive a transfusion from the donor?

Practice Exam 3

#1   We want to draw a sample of 5 without replacement from a population of 50 drivers convicted of speeding. If the drivers are labeled 01, 02, 03, ..., 50 and the following line is from a random number table

22368 46573 25595 85393 30995 89198 27982 53401 93965 34095 52666 19174

Which of the following represents the sample of 5, starting from the left?

(A)   22, 36, 8, 46, 32
(B)   22, 46, 25, 30, 27
(C)   22, 36, 25, 30, 27
(D)   22, 36, 46, 32, 39
(E)   22, 23, 36, 46, 32

#2   The histogram below was obtained from data from 300 basketball games in a junior high school basketball league. It represents the number of three-point baskets made in each game. A researcher takes an SRS of sample size n = 30 and computes the mean of each sample. Which of the following best describes the shape of the sampling distribution of the sample means?

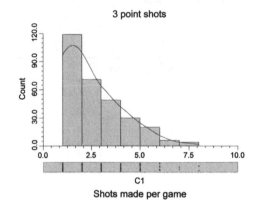

(A)   Skewed to the left
(B)   Skewed to the right
(C)   Uniformly distributed
(D)   Normally distributed
(E)   Cannot tell because the mean is unknown

#3   Thirty-five people from a random sample of 125 workers from Company A admitted to using sick leave when they weren't really ill. Seventeen employees from a random sample of 68 workers from Company B also admitted they had used sick leave when they weren't ill. What is the 95% confidence interval for the difference between the proportion of workers from the two companies who have misused sick leave?

(A) 0.03 ± 0.007
(B) 0.03 ± 0.008
(C) 0.03 ± 0.05
(D) 0.03 ± 0.11
(E) 0.03 ± 0.13

#4   The owners of an apple orchard want to estimate the mean weight of the apples they grow in order to plan for shipping more efficiently. The mean weight of a sample of nine apples is 6.3 ounces and comes from a population with a known standard deviation of 2.7 ounces. The orchard owners want to estimate the mean weight within 0.5 ounces at a 95% level. What is the minimum number of apples that need to be measured?

(A)   13
(B)   78
(C)   155
(D)   156
(E)   610

#5   A psychologist believes that students who prefer rough sports like football and hockey are more likely to have an active pet such as a dog. The table below shows data that were collected from a random sample of students with output from a statistical software program.

		Sport			Row Summary
		Basketball	Football	Hockey	
Pet	Cat	75	27	7	109
	Dog	115	61	10	156
	Fish	19	9	10	38
Column Summary		209	97	27	333

$\chi^2 = 21.23$ with df = 4. **WARNING: 1 of 9 cells have expected values less than 5**.

Which of the cells has an expected count less than 5?

(A)   Cat, hockey
(B)   Dog, hockey
(C)   Fish, basketball
(D)   Fish, football
(E)   Fish, hockey

#6   To increase the speed of typists, another keyboard layout called the DVORAK keyboard has been developed. Unlike the standard QWERTY keyboard layout, the home keys on the DVORAK keyboard are the most commonly used letters. Consequently, fingers do not have to travel as far to type passages. Which of the following designs would be most effective to test the hypothesis that the new keyboard layout improves typing speed?

(A)   Select 20 people who do not have good typing skills and randomly divide them into two equal-sized groups. Give each group six months of instruction and practice on the layout they are assigned to. At the end of six months have each person type the same long passage and determine the mean of each group.

(B)   Select 20 people who do not have good typing skills and let them choose which layout style they would like to learn. Give each group six months of instruction and practice on the layout they are assigned to. At the end of six months have each person type the same long passage and determine the mean of each group.

(C)   Select 20 people who have good typing skills and randomly divide them into two equal-sized groups. Give each group six months of instruction and practice on the layout they are assigned to. At the end of six months have each person type the same long passage and determine the mean of each group.

(D)   Select 20 people who have good typing skills and determine the mean typing speed for the group on a specific passage. Give each of the 20 people six months of instruction and practice on the new layout. At the end of six months have each person type the same long passage and determine the mean of the group and compare it to the mean for the earlier trial.

(E)   Select 20 people who have good typing skills and determine the mean typing speed for the group on a specific passage. Give each of the 20 people six months of instruction and practice on the new layout. At the end of six months have each person type the same long passage and determine the mean change for each individual.

#7   A poll showed that 60% of the adults in a large town are registered as Democrats. A newspaper reporter wants to interview a local Democrat regarding a recent decision by the City Council. If the reporter asks adults on the street at random, what is the probability that he will find a Democrat by the third person he stops?

(A)   0.936
(B)   0.216
(C)   0.144
(D)   0.096
(E)   0.064

#8   Students in an elementary class ran a 40 yard race and timed themselves. They then practiced sprinting for the next four weeks. The students again timed themselves. If the times were normally distributed, which of the following stem plots indicates an improvement in time.

(A)

First race		Last race
4	6	233
9974	6	5568
0233344	7	011223444
555778899	7	5667889
0014	8	11234
56779	8	67
		6|4 means 6.4 seconds

(B)

First race		Last race
4	6	02334
9974	6	55678
0233344	7	01234
555778899	7	56889
0014	8	11234
56779	8	55667
		6|4 means 6.4 seconds

(C)

First race		Last race
4	6	023344
9974	6	5589
0233344	7	024
555778899	7	56789
0014	8	02344
56779	8	5667789
		6|4 means 6.4 seconds

(D)

First race		Last race
4	6	233334
9974	6	55
0233344	7	011223444
555778899	7	569
0014	8	1123455
56779	8	677
		6|4 means 6.4 seconds

Choice E is on the next page.

(E)

First race		Last race
4	6	
9974	6	455689
0233344	7	011223444
555778899	7	5566678889
0014	8	112
56779	8	67
		6|4 means 6.4 seconds

#9   Box plots of two data sets are shown.

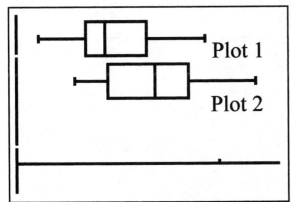

Based on the box plots, which of the following statements are true?

I.   The spread of both plots is similar.
II.   The means of both plots are similar.
III.   The sample sizes must be similar.

(A)   I only
(B)   II only
(C)   III only
(D)   I and II
(E)   II and III

#10   A biologist plotted data from an experiment. He was interested in obtaining a least squares regression line. However, the data did not appear to be linear, so a variety of transformations were applied. Based on the following residual plots, which transformation was most effective in linearizing the data?

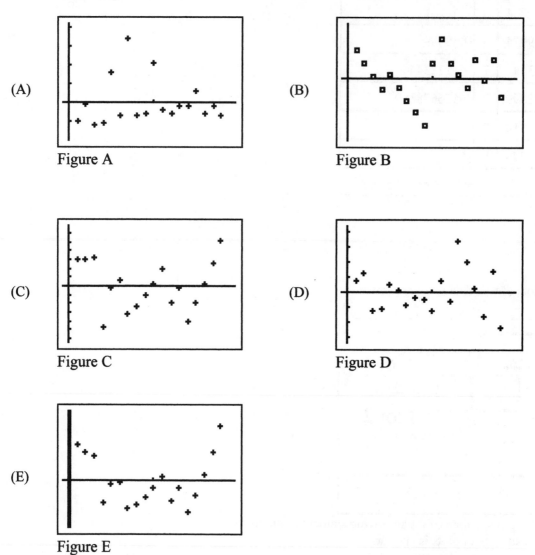

(A)
Figure A

(B)
Figure B

(C)
Figure C

(D)
Figure D

(E)
Figure E

#11   Researchers want to know if biofeedback will reduce the anxiety of people with claustrophobia. To investigate this idea, the researchers teach biofeedback techniques to 37 adult volunteers who have symptoms of claustrophobia. After six weeks of training 29 of the volunteers report a significant reduction in their anxiety. A statistician criticizes the experiment. Which of the following is likely to be the main criticism of the design?

(A)   People without claustrophobia were not included in the study.
(B)   No standardized measure of anxiety was described.
(C)   The volunteers were not divided into blocks that had different lengths of training.
(D)   No P-value was given for people to decide how significant the results were.
(E)   No control group was used to see if there was a placebo effect.

#12   A random sample of 65 female drivers from age 16-19 found the mean number of miles driven annually is 6780 with a standard deviation of 1545 miles. The 90% confidence interval is (6460, 7100). Which of the following statement is TRUE?

(A)   There is a 90% probability that the true mean is between 6460 and 7100 miles.
(B)   90% of all female drivers between 16-19 years old drive within 1545 miles of 6780 miles annually.
(C)   90% of all female drivers aged 16-19 drive at least 6100 miles annually.
(D)   If ten random samples of size $n = 65$ were taken from the population of interest, the true driving mean would be contained within the resulting interval about nine times.
(E)   None of the above statements is true.

#13   In the Mediterranean Sea 206 of 579 captured fish of a certain species were found to be infected by tapeworms. In the Atlantic Ocean fishers caught 132 fish of the same species and found that 28 were infected by tapeworms. Let p_1 equal the proportion of infected fish from the Mediterranean Sea. Let p_2 equal the proportion of infected fish from the Atlantic Ocean. What is the P-value of the hypothesis test: $H_0: p_1 = p_2$ vs. $H_a: p_1 \neq p_2$?

(A)   0.002
(B)   0.212
(C)   0.329
(D)   0.356
(E)   3.170

#14   An inspector checks large shipments of chickens to determine the proportion, p, in the shipment with unacceptable defects prior to processing the chickens. The inspector selects a random sample of 100 chickens from a shipment that is estimated to contain 2500 chickens. Only four of the chickens sampled had unacceptable defects. Which of the following conditions about a confidence interval of a proportion has not been met?

(A)   The data are an SRS from the population of interest.
(B)   The population is at least 10 times the sample size.
(C)   The sample size, n, is so large that both the count of successes, $n\hat{p}$, and the count of failures, $n(1-\hat{p})$, are at least 10.
(D)   The mean number of unacceptable chickens is normally distributed.
(E)   There are no apparent violations.

#15   A random variable X has a probability distribution as shown in the graph. What is the probability that X will be less than or equal to -0.60?

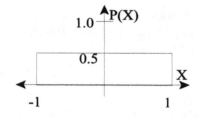

(A)   -0.60
(B)   -0.20
(C)   0.20
(D)   0.60
(E)   0.80

#16   A bike relay race is conducted with two person teams. One person rides on a level roadway and the other rides over hilly terrain and the times are combined. The mean time on the level roadway is 15 minutes with a standard deviation of 3 minutes. The mean time on the hilly course is 21 minutes with a standard deviation of 4 minutes. Assuming that the times from both portions of yhe race are normally distributed and the riders' times are independent of each other, what is the standard deviation of the combined time?

(A)   1 minute
(B)   5 minutes
(C)   6 minutes
(D)   7 minutes
(E)   Cannot be determined since the number of contestants is unknown

#17   A student computed a least squares regression line and found that the correlation coefficient was 0.83. In checking her answer she found she had switched the dependent and independent variables. She then computed the regression line using the correct order. What is the new correlation coefficient?

(A) $\dfrac{-1}{0.83}$

(B)   -0.83
(C)   0.689
(D)   0.83

(E) $\dfrac{1}{0.83}$

#18   An opinion research firm wants to find the country's reaction to a speech by a famous politician. To ensure a very narrow confidence interval, they decide to interview 6000 adults. The opinion company randomly selects six states and then selects ten postal Zip Codes at random from each state. Finally, 100 people from each Zip Code area are selected at random. This method of sampling is called

(A)   multi-stage sample.
(B)   simple random sample.
(C)   survey sample.
(D)   convenience sample.
(E)   stratified sample.

#19   In a study of 30 different energy bars a consumer advocacy group found that at least 60% of all bars contained more than 15 grams of carbohydrates. The manufacturer's spokesperson said the true proportion of bars with more than 15 grams of carbohydrates was no more than 20%. If p represents the proportion of bars with more than 15 grams of carbohydrates, which of the following gives the null and alternate hypotheses that the spokesperson is implying?

(A) $H_0: p = 0.60$
        $H_a: p < 0.60$
(B) $H_0: p = 0.20$
        $H_a: p > 0.20$
(C) $H_0: p > 15$
        $H_a: p \leq 15$
(D) $H_0: p \leq 0.60$
        $H_a: p > 0.60$
(E) $H_0: p < 0.20$
        $H_a: p = 0.20$

#20   A χ^2 analysis was done on the following data that compared income with educational level.

Income	$0 - 20,000	$20,001- 40,000	$40,001 - 60,000	$60,001 - 80,000	$80,001 +	Row summary
Graduate degree	12	23	42	45	18	140
Bachelor's degree	15	25	51	41	18	150
High school graduate	32	38	31	17	12	130
High school drop out	17	15	10	8	7	57
Column summary	76	101	134	111	55	477

What is the probability that the person has earned a Bachelor's degree given that the income level was between $60,001 and $80,000?

(A)   Less than 0.001
(B)   0.09
(C)   0.27
(D)   0.37
(E)   42.14

#21   A physician can use one of two drugs to treat a certain disease. Drug A will cure 75% of the cases but has unwanted side effects in 40% of the cases. Drug B will cure 50% of the cases and has unwanted side effects in 30%. Sixty patients are given Drug A and 80 patients receive Drug B. If each patient receives only one drug, what is the expected number of patients who will experience unwanted side effects?

(A)   30
(B)   48
(C)   49
(D)   85
(E)   88

#22 An experiment was done to determine the time it takes for a mouse to run through a maze. The results are summarized in the box plots below. The top plot shows the times by mice that were 15 days old. The bottom plot shows the times by mice that were 20 days old. Based on the information from the box plots, which of the following statements is correct?

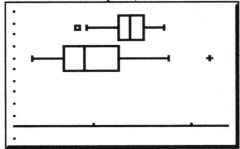

Times to run through a maze

(A) About 60% of the older mice had slower times than any of the younger mice.
(B) The standard deviation of the younger mice is higher than the standard deviation of the older mice.
(C) There were the same number of mice in each trial.
(D) The times for the older mice would have a narrower 95% confidence interval than the younger mice.
(E) About 75% of the older mice had faster times than all but 25% of the younger mice.

#23 A police department in a large city tallies the number of traffic tickets written by the department's 478 officers during the month of December. The chief wants to determine the 95% confidence interval for the mean number of tickets written by each officer. Which of the following is the reason why the chief cannot compute the confidence interval?

(A) December is not a typical month and is therefore not representative of the entire year.
(B) There are 477 degrees of freedom, which cannot be found in the table exactly.
(C) It isn't known if the data are normally distributed.
(D) A confidence interval cannot be constructed when population data are known.
(E) If a few officers wrote an inordinate number of tickets, they could skew the data.

#24 A pharmaceutical company developed a new medicine that eliminated joint pain among arthritis sufferers. When the company tested it against a similar medication from another company, a sample size of 100 users in a double blind test showed no statistical difference between the two medicines. However, when the sample size was increased to 5000 users, the results showed a clear, statistical difference that the new medication was more effective. The first study illustrates the principle of

(A) stratified sampling.
(B) t-test vs. z-test.
(C) type I error.
(D) low power.
(E) strong replication.

#25   According to the Piano Manufacturers Association, 1800 new pianos were sold in the Salt Lake City metropolitan area in 2000. In the Chicago metropolitan area 1900 new pianos were sold in the year 2000. The population of the Salt Lake City area is 1.5 million and the population of the Chicago area is 6 million. If the proportion of piano purchases were the same for both cities, what is the probability of getting a result at least this extreme?

(A)   Less than 0.0005
(B)   0.0005
(C)   0.0012
(D)   0.9988
(E)   43.576

#26   A fire department in a rural county reports the mean response time is 22 minutes. A home owner was told the response time of 30 minutes to his neighborhood was at the 3rd quartile. What standard deviation did the report use if the times were normally distributed?

(A)   2.7 minutes
(B)   3.4 minutes
(C)   7.1 minutes
(D)   11.9 minutes
(E)   16.0 minutes

#27   A study was done among recent graduates from a business school in Texas comparing the length of time to be hired between Hispanics and non-Hispanics. For each graduate three variables were measured: ethnic background, area of specialty, and length of time until hired. Which of the following statements is correct?

(A)   Ethnic background and area of specialty are explanatory variables; hiring time is a response variable.
(B)   Ethnic background and hiring time are explanatory variables; area of specialty is a response variable.
(C)   Ethnic background is an explanatory variable; hiring time is a response variable; area of specialty is a possible confounding variable.
(D)   Hiring time is the explanatory variable; area of specialty and ethnic background are response variables.
(E)   Hiring time is an explanatory variable; ethnic background is a response variable; area of specialty is a possible confounding variable.

#28   Twenty-five people were asked to participate in an experiment to determine the side effects of a new medication. Each person received a dose of medicine with a dye and one dose without the dye on another day. The decision as to which dose was given first was decided by flipping a coin. Which of the following is the best reason for using a paired t-test instead of a two-sample t-test?

(A)   Since each person received more than one treatment, fewer subjects were needed.
(B)   Lurking variables can be eliminated since we can compare both treatments on the same subject.
(C)   The test statistic is more accurate since we work only with the differences.
(D)   Fewer assumptions are required since the desired parameter is the difference.
(E)   Each person can be judged on two treatments.

#29   The speed of a typist is in the top 1% of all typists tested by a State Employment Service. If the person's typing speed is 120 wpm and the population standard deviation is 24 wpm, which of the following is closest to the mean typing speed for all people tested, assuming the typing speeds aer normally distributed?

(A)   46 wpm
(B)   60 wpm
(C)   64 wpm
(D)   71 wpm
(E)   72 wpm

#30   A psychologist theorizes there was a strong relationship between a person's IQ score and their sense of humor. She randomly selected 45 adults and found the following results.
IQ: Mean = 105. Standard deviation = 12.
Durante Test of Relevant Humor: Mean = 140. Standard deviation = 24.
$r^2 = 0.81$.
What is the predicted humor score if the IQ score of an individual is 110?

(A)   51
(B)   142
(C)   148
(D)   149
(E)   Cannot be determined since the individual data points are not known

#31   An eye clinic has two doctors. Dr. Smith recommends glasses 60% of the time and corrective surgery 40% of the time. Dr. Wilson recommends glasses 70% of the time and surgery 30% of the time. 60% of all patients are treated by Dr. Wilson. If a patient from the clinic is selected at random, what is the probability that surgery was recommended?

(A)   18%
(B)   34%
(C)   35%
(D)   58%
(E)   70%

#32   The graph of a data set is shown below.

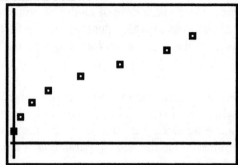

Which of the following transformations would be appropriate to linearize the data for a regression analysis?

I. $\sqrt{y}$

II. y^2

III. $\ln y$

(A)   I only
(B)   II only
(C)   III only
(D)   I and II
(E)   I and III

#33   In a large city, there are three newspapers published, *The Post, The Gazette,* and *The Tribune*. 45% of the homes subscribe to *The Post* and 40% subscribe to *The Gazette*, while 25% subscribe to both *The Post* and *The Gazette* only. If a home is randomly selected, what is the probability that it subscribes to *The Tribune*?

(A)   4.5%
(B)   15%
(C)   18%
(D)   40%
(E)   60%

#34   According to The Insurance Institute for Highway Safety, the national accident rate for 16- year-old males was 210 accidents for every 1000 drivers during the year 2000. In a large city in the Midwest a random sample of 150 16-year-old male drivers found that 36 of them had been in an accident during the year 2000. What is the probability that the accident rate of 16-year-old male drivers in this area is greater than the national rate?

(A)   0
(B)   0.184
(C)   0.367
(D)   0.882
(E)   0.902

#35   A veterinarian sampled 235 sheep from a large herd to obtain a 90% confidence interval for the mean weight of wool from each sheep. The interval was (23.7 lbs., 33.6 lbs). If the veterinarian had used a 95% confidence interval instead, the confidence interval would be

(A)   wider and would have a smaller risk of missing the true mean.
(B)   narrower and would have a larger risk of missing the true mean.
(C)   narrower and would have a smaller risk of missing the true mean.
(D)   wider and would have a larger risk of missing the true mean.
(E)   wider but the risk of missing the true mean cannot be determined.

#36   A χ^2 test was performed on a two-way table of values. The value of the χ^2 statistic was 10.5. Every value in the two-way table was multiplied by 4. The value of the new χ^2 statistic is which of the following?

(A)   10.5
(B)   14.85
(C)   21.0
(D)   42.0
(E)   Cannot be determined without knowing the values of the original table

#37   The dot plot below represents the lengths (in cm) of 150 fingerlings that are six weeks old in a fish hatchery.

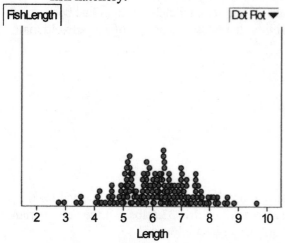

Which of the following is a normal percentile plot of the same data?

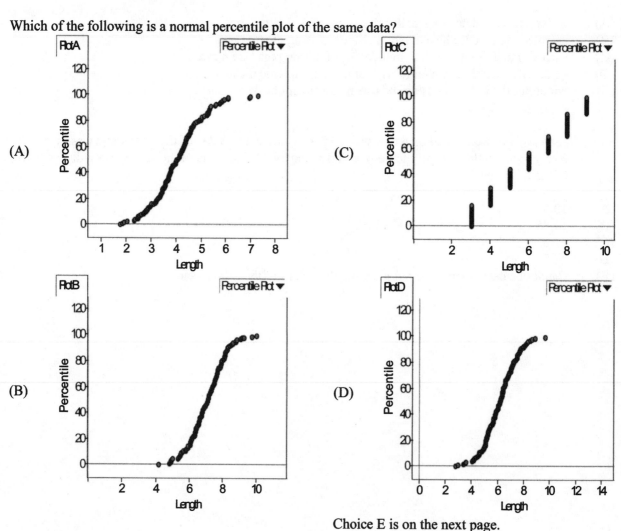

Choice E is on the next page.

(E)

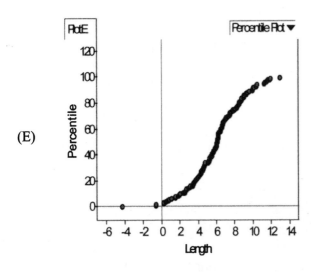

#38 A materials engineer found a linear relation between the weight hung from a wire and the distance the wire stretched. The linear regression model is
$\hat{y} = 1.2 + 1.5x$ for $1 \leq x \leq 10$,
where x is the weight in kg and $\hat{y}$ is the estimated length of the wire in mm. What is the estimated increase in length (in mm) that corresponds to an increase in weight of 4 kg?

(A) 1.5
(B) 2.7
(C) 6.0
(D) 7.2
(E) 16.2

#39 A basketball team with 11 players of differing heights trades its shortest player to another team for a tall center who is now the tallest player on the team. Which of the following statements is FALSE?

(A) The mean height of the team will increase.
(B) The median height will stay the same.
(C) The standard deviation might be different.
(D) The mode will stay the same.
(E) None of the above

#40   A tax assessor developed a model that predicted the value (in thousands of dollars) of a home based on the square footage of the home. The accompanying computer printout summarizes the findings.

Dependent variable: HOMEVALU

Analysis of Variance

Source	DF	Sum of Squares	Mean Square	F Value	Prob>F
Model	1	17.33721	17.33721	14.043	0.0027
Error	12	13.66329	2.71277		
C Total	13	31.0005			

Root MSE		1.23758	R-square		0.6493
Dep Mean		1.43754	Adj R-sq		0.5962
C.V.		72.88631			

Parameter Estimates

Variable	DF	Parameter Estimate	Standard Error	T for H0: Parameter = 0	Prob > |T|
INTERCEP	1	35.70032	6.9533214	13.049	0.0237
SQFEET	1	0.02875	0.0030725	4.014	0.0017

Based on the information, which of the following is closest to the predicted value of a home with 2000 square feet?

(A) $93,200
(B) $129,500
(C) $130,000
(D) $227,000
(E) $804,000

#1

A police department used a radar gun to measure the speed of a sample of cars on a residential street. The results are shown in the are summarized in the table below

Speed (mph)	21	22	23	24	25	26	27	28	29	30	31	32	33	34	35
Number of cars	2	4	5	10	15	21	43	55	38	29	15	9	3	0	2

(A)   Create a frequency histogram of the data. Let the interval width be 2 miles/hour.

(B)   Give the five number summary of the data.

(C)   Using the information in parts (A) and (B), describe the shape of the data.

(D)   Traffic engineers set the speed limit at the 85th percentile of the traffic load. Based on the data, compute the recommended speed limit (to the nearest mile per hour) for this road.

#2

A gasoline company has developed a new additive for gasoline that is supposed to improve the gas mileage of a car. A consumer organization has been asked to determine if the additive is effective. They have ten professional drivers and ten six-passenger cars of the same make and model and ten four-passenger cars of the same make and model.

(A)   Design an experiment to test the effectiveness of the additive on both the six-passenger cars and the four-passenger cars.

(B)   Explain what type of hypothesis test you would use to test the effectiveness of the additive. Write both the null and alternate hypotheses. Define any variables used.

Data from the high jump and long jump events during the Summer Olympics from 1896 through 1984 is shown in the table. (Year 1900 = 0). The numbers are the winning jumps (in inches) for that year.

Year	High jump	Long jump	Year	High jump	Long jump	Year	High jump	Long jump	Year	High jump	Long jump
-4	71.2	249.8	20	76.2	281.5	48	78	308	68	88.2	350.5
0	74.8	282.9	24	78	293.1	52	80.3	298	72	87.8	324.5
4	71	289	28	76.4	304.8	56	83.2	308.2	76	88.5	328.5
8	75	294.5	32	77.6	300.8	60	85	319.8	80	92.8	336.2
12	76	299.2	36	79.9	317.3	64	85.8	317.8	84	92.5	336.2

Two scatter plots are shown below

$HighJump = 71.8 + 0.223 * Year$

$r^2 = 0.92$

A scatter plot of the data, the least squares regression line, and the residual plot are shown in the figure below.

$LongJump = 60 + 3.06 * HighJump$

$r^2 = 0.78$

A scatter plot of the data, the least squares regression line, and the residual plot are shown in the figure below.

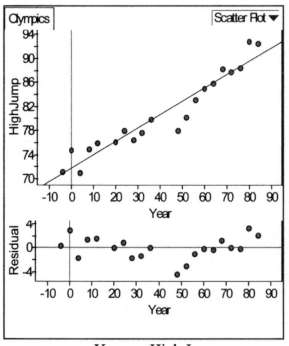

Year vs. High Jump

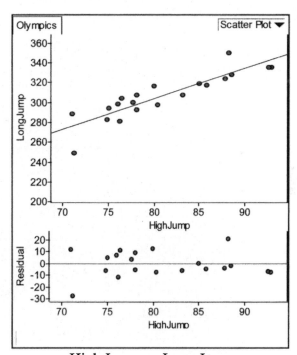

High Jump vs. Long Jump

(A)   Consider the information for Year vs. High Jump. Is the model proposed by the linear regression line a reasonable model? Explain.

(B)   For the graph Year vs. High Jump the residual with the largest absolute value occurred in 1948. What is the residual value for this year? Show your work. Explain what it means in terms of the model.

(C)   If the winning High Jump was 90 inches, what would be the predicted length of the winning Long Jump?

(D)   The value of r^2 for High Jump vs. Long Jump is 0.78, which indicates a strong relationship. Is this relationship causal or associative? Why?

#4

In a recent study, researchers evaluated the gender patterns of 4208 families with exactly 2 children. The results are summarized below

Pattern   Frequency
Boy-Boy   1085
Boy-Girl   1086
Girl-Boy   1111
Girl-Girl   926

(A)   Given that the first child born in a two-child family is a girl, what is the probability that the second child will be a boy?

(B)   Test the claim that if the first child in a two-child family is a girl, there is a higher proportion of boys born as the second child.

#5

In some countries there are no primary elections to narrow down the field of candidates, so several candidates could run for a particular office. If a candidate does not receive a majority, then the two candidates with the most votes run against each other in a runoff election a few weeks later. Suppose that three candidates - Candidate A, Candidate B, and Candidate C -- are running from three different regions of the country -- Region 1, Region 2, and Region 3.

Candidate A is from Region 1. Based on previous results, Region 1 casts about 35% of all votes in the country. Polls indicate that Region 1 will cast about 70% of its votes for Candidate A, 20% for Candidate B, and 10% for Candidate C.

Candidate B is from Region 2. Based on previous results, Region 2 casts about 50% of all votes in the country. Polls indicate that Regions 2 will cast about 20% of its votes for Candidate A, 60% for Candidate B, and 20% for Candidate C.

Candidate C is from Region 3. Based on previous results, Region 3 casts about 15% of all votes in the country. Polls indicate that Region 3 will cast about 25% of its votes for Candidate A, 15% for Candidate B, and 60% for Candidate C.

(A)   If the actual vote is in line with the preliminary estimates, will one of the candidates receive a majority of the votes? If not, which two candidates will be in the runoff election? Show your work that leads to your conclusion.

(B)   What proportion of her total vote did Candidate C receive from her home region?

#6

The Wonderful Chocolate Chip Cookie Company promises at least three chocolate chips in each cookie. They promise a double your money back guarantee if a cookie doesn't have at least three chips. We want to simulate the number of chips that must be added to the dough for a batch of six cookies to make sure that each cookie has at least three chips.

(A)   Draw six small circles to represent six cookies. Number them 1-6. Use the following random number table to simulate the adding of chocolate chips to the batch. If the random number is 1, then cookie #1 gets a chip, etc. Continue adding chips until each cookie has at least three chips in it. Tally the number of chips that were added to the batch. Clearly indicate which numbers you are using.

Perform the simulation a second time.

20594	69574	13353	67267	39205	07136	93007	67214	91790	01273
67456	42320	60053	59799	63251	82094	64425	92476	20756	16082
99437	40160	38150	81329	85035	58152	96571	24826	41897	43810

(B)   Suppose that the simulation was performed 30 times. The mean number of chips added until each cookie had at least three chips was 26.3 with a standard deviation of 4.8 chips. The distribution was approximately normal with no outliers.

Construct a 95% confidence interval for the number of chips necessary to add to a batch of dough to get three chips per cookie.

(C)   Is it possible to guarantee that *every* cookie will have at least three chocolate chips? Why?

(D)   Suppose the company decides it only wants to pay off on the money back guarantee 5% of the time. What is the least number of chips that should be added to a batch of dough for six cookies to accomplish this?

Practice Exam 4

#1   In a large city, there are three newspapers published, *The Post, The Gazette,* and *The Tribune.*
45% of the homes subscribe to *The Post*, 40% subscribe to *The Gazette*, and 25% subscribe to
both *The Post* and *The Gazette.* If a home is randomly selected, what is the probability that it
subscribes to either *The Post* or *The Gazette* but not both?

(A)   15%
(B)   20%
(C)   35%
(D)   40%
(E)   75%

#2   The graph below represents the water content in a watershed snow pack during a recent winter
from October through April. There was 200 total inches of snow with a total water content of 11
inches. The residual plot for a least squares regression line is plotted immediately below the
original plot.

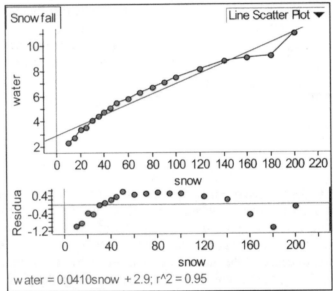

Which of the following statements is TRUE?
   I.   A linear regression is a good model.
   II.   The residual for a snowfall of 100 inches has a positive value.
   III. Approximately half the water content was realized after about half the snowfall.

(A)   I only
(B)   II only
(C)   III only
(D)   I and II
(E)   I and III

93

#3   A researcher wants to determine the mean income of adults in a particular state. She decides to take a random sample of 1525 citizens with a driver's license. Which of the following statements is NOT correct?

(A)   People without a license will be missed in the sample.
(B)   People from another state will not be included.
(C)   The mean income of the sample will be reasonably close to the mean of the population.
(D)   It is likely that most people of driving age will be included.
(E)   The sample standard deviation will be similar in value to the actual standard deviation.

#4   The histogram below represents the number of defective headlights in 250 separate shipments. A quality control analyst takes an SRS of sample size $n = 25$ and computes the mean of each sample. Which of the following best describes the shape of the sampling distribution of the sample means?

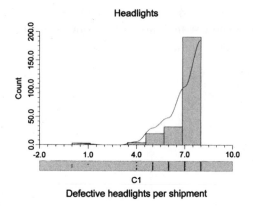

Defective headlights per shipment

(A)   Skewed left with a mean of approximately 7
(B)   Skewed left with a mean of approximately 4
(C)   Normally distributed with a mean of approximately 7
(D)   Normally distributed with a mean of approximately 4
(E)   Cannot be determined because the standard deviation is unknown

#5   In an experiment students place one hand into a container full of pennies and try to grab as many as they can. The number of coins grabbed by boys has a normal distribution with a mean of 32 coins and a standard deviation of 6 coins. The number of coins grabbed by girls is also normally distributed with a mean of 23 coins and a standard deviation of 4 coins. One boy and one girl each grab a handful of coins. If the number of coins drawn by each is independent, what is the probability that the boy grabs no more than five coins more than the girl?

(A)   0.21
(B)   0.29
(C)   0.34
(D)   0.55
(E)   0.58

#6   In a recent study, a random sample of 250 newly-planted trees along the Oregon coast showed a significant positive relationship between the distance from the shoreline and the height of the tree. The distance from the shoreline is

(A)   a block design with the distance from the shoreline as the block.
(B)   a response variable.
(C)   a confounding variable.
(D)   an explanatory variable.
(E)   a random variation.

#7   Which of the following is NOT a condition for a chi-square test?

(A)   The data must come from a simple random sample.
(B)   No more than 20% of the cells can have an expected value less than 5.
(C)   The data must be independent.
(D)   All expected cell counts must be at least 1.
(E)   The number of rows must be less than the number of columns.

#8   A random sample of tree heights in an apricot orchard is summarized in the computer output below.

DESCRIBE 'HEIGHT'

	N	MEAN	MEDIAN	TRMEAN	STDEV	SEMEAN
Height	40	20.125	19.000	20.194	4.592	0.726

	MIN	MAX	Q1	Q3
Height	10.000	29.000	17.250	23.000

Which of the following five stem and leaf plots is most likely to be the data set?

(A)
```
Stem-and-leaf of PlotA   N = 40
Leaf Unit = 1.0

   1 01
   1 3
   1 4
   1 666777
   1 88888888999
   2 0001
   2 333333
   2 45
   2  66677
   2  89
```

(C)
```
Stem-and-leaf of PlotC    N = 40
Leaf Unit = 1.0

   1 0001
   1 2333
   1 4445
   1 6677
   1 88888999
   2 0111
   2 333
   2 455
   2  667
   2  899
```

(B)
```
Stem-and-leaf of PlotB    N = 40
Leaf Unit = 1.0

   1 00011
   1 23
   1 4445
   1 666777
   1 88888999
   2 01
   2 3333
   2 4
   2  6677
   2  8899
```

(D)
```
Stem-and-leaf of PlotD    N = 40
Leaf Unit = 1.0

   1 0011
   1 233
   1 4445
   1 666777
   1 8888899999
   2 011
   2 333
   2 4
   2  667
   2  889
```

Choice E is on next page.

(E)
```
Stem-and-leaf of PlotE  N  = 40
Leaf Unit = 1.0

        1  00011
        1  23
        1  44
        1  666777
        1  88888999
        2  01
        2  3333
        2  4
        2  6677
        2  8899
        3  01
```

#9   In 1970 the U.S. Armed Services instituted a lottery as a random selection process when 19-year-old males were drafted into service. Each of the 366 possible birth dates was placed in a separate capsule and each capsule was placed in a large plastic drum. The box plots for draft eligible numbers for each month is shown. Those with the lowest draft lottery numbers were drafted before others.

1970 Draft Lottery

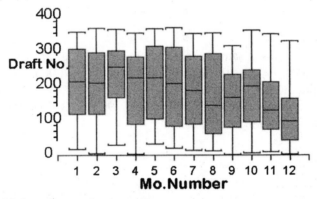

Based on the box plots, which statement is correct?

(A)   There seems to be no relationship between birth month and draft lottery number.
(B)   There is a weak negative correlation between birth day and draft lottery number.
(C)   Since the spread for each month is about the same, the mean draft lottery number for each month is about the same.
(D)   Those people born near the end of the year tended to be assigned lower draft lottery numbers than those born at the first of the year.
(E)   Since there is a region common to all box plots, the confidence intervals for each month all cover a similar value for the mean draft lottery number.

#10   A study done by the Duke University Medical School compared the effectiveness of St. John's

Wort as a natural alternative to reducing depression. In a double blind study 100 people with mild to moderate depression were given daily doses of St. John's Wort. Another 100 people with similar symptoms were given a sugar pill. Assignments were made randomly. Each participant was asked to count the number of days they exhibited feelings of depression. A two sample t-test on the difference in the mean number of days of depression was performed. The P-value was 0.42. This result can best be explained by saying

(A)   about 42% of the time a sample drawn from populations with no differences would show a difference as extreme or more so as those found in the study.
(B)   42% of the participants did not experience any relief.
(C)   there was a 42% drop in the mean number of days of depression in the control group.
(D)   there was a 42% drop in the mean number of days of depression in the group that received the St. John's Wort.
(E)   the was a 42% drop in the difference in the number of days of depression between the two groups.

#11   A company is looking for golfers who can hit the ball at least 300 yards. Suppose the probability at a certain golf course that one of the players can hit a golf ball 300 yards is 0.35. If six people are selected at random, what is the probability that at least four of them can hit a golf ball at least 300 yards?

(A)   0.012
(B)   0.022
(C)   0.095
(D)   0.117
(E)   0.883

#12   A researcher wants to draw a sample of 5 from a population of 261 business days from the New York Stock Exchange. Suppose the days are labeled 001, 002, 003,..., 261 and the following line is taken from a random number table.
28918 69578 88231 33276 70997 79936 56865 05859 90106 31595 01547 85590
Starting at the left, what is the sample of 5?

(A)   186, 231, 132, 10, 159
(B)   28, 91, 86, 95, 78
(C)   28, 169, 88, 133, 70
(D)   186, 231, 99, 106, 154
(E)   28, 69, 88, 33, 70

#13   In a recent study, a random sample of tomato plants showed a significant negative relationship between the acidic content of the soil and the number of tomatoes produced. The amount of acid in the soil is

(A)   a controlled factor.
(B)   an explanatory variable.
(C)   the statistical result of a high correlation coefficient.
(D)   a response variable.
(E)   a confounding variable.

#14   A cereal maker claims that the weight of its new packages is 48 ounces. An investigator from the Division of Weights and Measures believes the packages hold less than the stated amount. Let μ be the true mean weight of all packages produced by the company. Which of the following statements represents the null and alternate hypotheses the investigator should test?

(A) $H_0: \mu = 48$ oz.
       $H_a: \mu < 48$ oz
(B) $H_0: \mu > 48$ oz.
       $H_a: \mu < 48$ oz
(C) $H_0: \mu = 48$ oz.
       $H_a: \mu \neq 48$ oz
(D) $H_0: \mu < 48$ oz.
       $H_a: \mu > 48$ oz
(E) $H_0: \mu > 48$ oz.
       $H_a: \mu \leq 48$ oz

#15   A reporter believed that police officers were required to write a specific quota of traffic tickets during a month. In order to meet the alleged quota, he believed officers would need to write more tickets during the last week of the month. To investigate the claim, the reporter collected the number of tickets written by the local police force in a month and organized them by weeks as shown in the table below.

Week	First week	Second week	Third week	Fourth week	Total
Tickets written	133	112	154	165	564

A chi-square analysis was performed to test the claim that there is a relationship between the week of the month and the number of tickets written. What is the P-value of the test?

(A) $0.0005 < P < 0.001$
(B) $0.0025 < P < 0.005$
(C) $0.005 < P < 0.01$
(D) $0.01 < P < 0.02$
(E) $0.025 < P < 0.05$

#16　A study was done comparing the miles obtained by tires on cars with automatic transmissions versus cars with standard transmissions. Ten identical cars with automatic transmissions were all fitted with new tires from the same manufacturer. Ten other cars that were identical to the first, except that they had standard transmissions, were fitted with the same brand of new tires. Professional drivers drove over the same course at identical speeds. The results from a computer analysis are shown below. Let C1 represent cars with automatic transmissions and C2 represent cars with manual transmissions.

TWO SAMPLE T FOR C1 VS C2				
	N	MEAN	STDEV	SE MEAN
C1	10	29930	1735	549
C2	10	29530	1527	483
95 PCT CI FOR MU C1 - MU C2: (-1136, 1936)				
TTEST MU C1 - MU C2 (VS NE): T= 0.55 P = 0.59 DF = 18				
POOLED ST DEV = 1634				

Based on the information, which of the following statements is TRUE?

(A)　95% percent of the time, the confidence interval will contain 0.
(B)　A matched pair t-test should have been used.
(C)　The population mean of tire mileage on cars with automatic transmissions is higher than tire mileage from cars with manual transmission.
(D)　 Statistically, there is no difference between the population means of the two samples.
(E)　The sample size is too small to draw any conclusion.

#17   A histogram of a data set is shown below

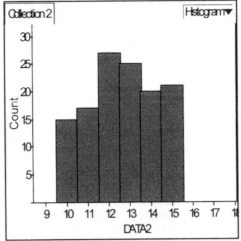

Which of the following five figures represents a box plot of the same data?

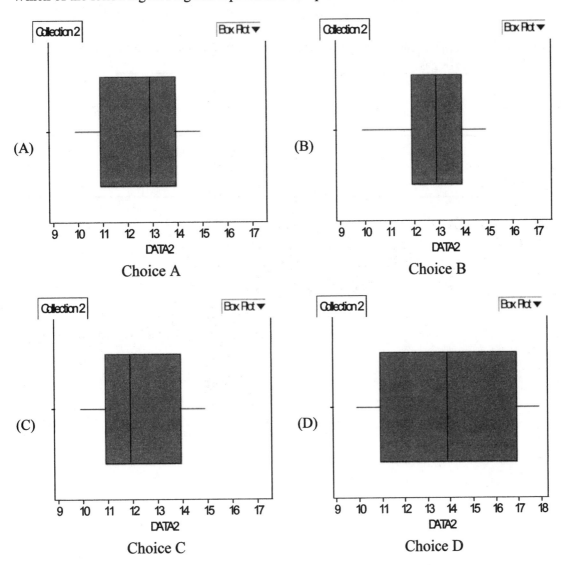

Choice A   Choice B

Choice C   Choice D

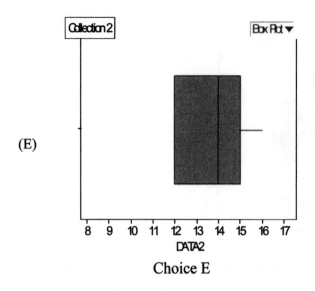

(E)

Choice E

#18   The table below gives the number of times a certain number of gallons of gasoline were purchased at a self-serve gas station.

Number of gallons of gasoline	6	7	8	9	10	11	12	13
Number of purchases (n = 75)	5	11	4	10	12	11	14	8

What is the value of the InterQuartile Range?

(A)   3
(B)   4
(C)   5
(D)   6
(E)   9

102

#19   A hospital administrator wants to determine a 94% confidence interval for the mean number of days that a patient stays at the hospital. The data given her by the business manager list the stay for each patient for the previous calendar year. Which of the following is the reason the administrator cannot determine the confidence interval?

(A)   A confidence interval cannot be constructed when population data are known.
(B)   Since most stays are just a few days long, the data are highly skewed.
(C)   A 90% or 95% confidence interval can be computed but not a 94% confidence interval since the tables don't show the critical values.
(D)   Since many patients were treated in the Emergency Room and then released, their stay would be zero days. This cannot be factored into the confidence interval.
(E)   Some patients may have been treated at the hospital more than once, which invalidates the independence requirement for a confidence interval.

#20   Scientists wanted to develop an insecticide that was harmless to humans. They tested their product on rats and mice. The data seemed to indicate that, in each trial, a greater proportion of rats experienced negative side effects than did the mice. They repeated the experiment 83 times and recorded the difference in the proportions. As part of the inference test, they plotted the differences. Which of the following describes the shape of the dot plot?

(A)   Approximately normal
(B)   Randomly scattered with no apparent pattern
(C)   Bi-modal
(D)   Uniformly distributed
(E)   Skewed strongly to one direction

#21   In 1999 2.2% of all cars in the United States were reported stolen. In a random sample of 400 Nissan Maxima cars 12 were reported stolen during 1999. What is the probability that the theft rate of the Nissan Maxima is greater than the national rate for all cars.

(A)   0.14
(B)   0.17
(C)   0.28
(D)   0.35
(E)   0.48

#22 The results of a random sample of ten paired observations is shown in the table below.

TEST OF MU = 0.000 VS MU LESS THAN 0.000 POPULATION1 - POPULATION2						
	N	MEAN	STDEV	SE MEAN	T	P-VALUE
DIFF	15	-4.000	4.1747	1.0779	-3.71	0.00116

	N	MEAN	STDEV	SE MEAN	95.0 PERCENT C.I.	
DIFF	15	-4.00	4.1747	1.0779	(-6.312,	-1.688)

Based on the output which of the following statements is FALSE?

(A)   It is unlikely that the difference between each pair of observations is equal to 0.
(B)   Each member of Population1 is less than its paired member from Population2.
(C)   There are 14 degrees of freedom.
(D)   t-procedures can be performed on data from 15 observations.
(E)   The SE MEAN = $\dfrac{STDEV}{\sqrt{15}}$.

#23   The residual plot of a small data set is shown. Which point is most likely an influential observation?

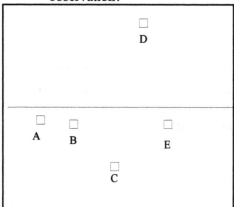

(A)   A
(B)   B
(C)   C
(D)   D
(E)   E

#24   The U.S. Post Office uses scanners to read and sort mail according to the bar codes of Zip Codes printed on letters. The mean speed of a postal scanner is 2000 letters per hour with a standard deviation of 150 letters per hour. If the speed of a scanner falls below the 30th percentile, it is sent out for repairs. What rate will cause a scanner to be sent out, assuming the scanning speeds are normally distributed?

(A)   1400
(B)   1718
(C)   1922
(D)   1943
(E)   1955

#25   Golf courses have a wide range of difficulty. Similarly, players differ in ability. In order to adjust for variations between players, they are often assigned a handicap score. To adjust for variations between courses, a handicapper decides to compare the golfer's score against the data from the course. Suppose course A plays at an mean score of 76 with a standard deviation of 8 strokes with a normal distribution of scores. The mean score for course B is 80 with a standard deviation of 6 stokes and the scores are normally distributed. If a golfer regularly shoots an 80 on course A, what should be the comparable score on course B?

(A)   80
(B)   83
(C)   84
(D)   86
(E)   88

#26   Nine people are randomly selected from a group of swimmers. They are asked to hold their breath for as long as possible and the times are recorded. The nine people are then instructed in a new method to hold their breath. Afterwards, they are asked to hold their breath and the times are again recorded. The results are shown in the following table:

Person	1	2	3	4	5	6	7	8	9
1st trial (in seconds)	67	68	73	77	69	65	70	69	70
2nd trial (In seconds)	69	68	74	77	72	68	71	72	71

The null hypothesis H_0: *The training had no effect on the length of time a person could hold his/her breath* was tested vs. the alternate hypothesis H_a: *The training increased the time each person was able to hold his/her breath.* Using the proper test statistic, which is the correct decision (let $\alpha = 0.05$).

(A)   Reject H_0 if the test statistic is greater than 1.645.
(B)   Reject H_0 if the test statistic is greater than 1.812.
(C)   Reject H_0 if the test statistic is greater than 1.833.
(D)   Reject H_0 if the test statistic is greater than 1.860.
(E)   Reject H_0 if the test statistic is greater than 2.262.

$\hat{y} = 7.2 + 3.6x$

$\bar{x} = 1.5, s_x = 2$

$\bar{y} = 12.6, s_y = 8$

What is the value of r^2?

(A)   0.12
(B)   0.25
(C)   0.81
(D)   0.90
(E)   1.23

#28   A random variable X has a probability distribution as shown in the graph below. What is the probability that X will be greater than or equal to 0.5?

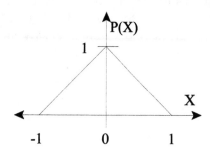

(A)   0.87
(B)   0.75
(C)   0.50
(D)   0.25
(E)   0.13

#29   Which of the following box plots represents a Normal Distribution?

(A)

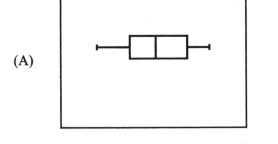

(C)

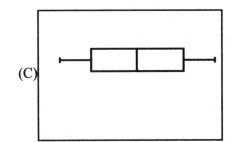

(B)

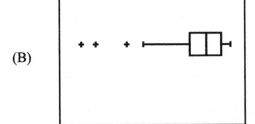

(D)

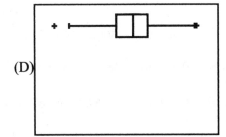

(E)

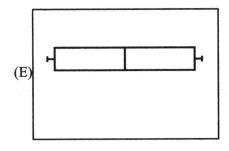

#30   A dice game pays a player $5 for rolling a 3 or a 5 with a single die. The player has to pay $2 for any other roll. If a person plays the game 30 times, what is the probability that the person will win at least $15.

(A)   0.0030
(B)   0.0643
(C)   0.2767
(D)   0.3085
(E)   0.3897

#31   A car rental agency has two locations, one by the airport and the other downtown. The agency rents 70% of its cars at the airport and the rest downtown. At the airport 20% of the rentals are Sport Utility Vehicles. At the downtown location, 40% are SUVs. If a customer is selected at random, what is the probability that she did NOT rent an SUV?

(A)   26%
(B)   40%
(C)   66%
(D)   70%
(E)   74%

#32   A confidence interval is robust when

(A)   changes in the required level of confidence, 1- α, do not affect the width of the confidence interval.
(B)   the confidence interval does not change if the required assumptions for the procedure are violated.
(C)   the confidence interval does not change very much if the required assumptions for the procedure are violated.
(D)   the confidence interval is at least 2 standard deviations wide.
(E)   the Standard Error of the Mean is least than 1.0.

#33   Three red balls, four white balls, and one green ball are in a bag. A ball is drawn without replacement. What is the probability that three balls can be drawn without drawing the green ball?

(A)   0.179
(B)   0.625
(C)   0.643
(D)   0.670
(E)   0.998

#34   Runners competed in a local road race. The mean finishing time for the race was 43.5 minutes with a standard deviation of 16.2 minutes. The sponsors wanted to have a special race for those who were in the fastest 10%. Assuming the times were normally distributed, which of the following is the cutoff time?

(A)   22.8 minutes
(B)   25.7 minutes
(C)   39.2 minutes
(D)   42.2 minutes
(E)   64.3 minutes

#35   Students in an AP Statistics class wanted to determine the percentage of students in their school who walked to school each day. There are 2230 students in the school. The student directory is numbered from 1-2230. The class selects every 10th student from the school for a total of 223 students to poll. What is this type of sampling called?

(A)   Multi-stage sample
(B)   Simple random sample
(C)   Survey sample
(D)   Systematic sample
(E)   Stratified sample

#36   In an aquarium are 36 guppies of a variety of sizes and ages. The length of each guppy is measured and recorded. A researcher drops a dividing wall down the middle of the aquarium, trapping half the guppies on each side. One side gets an extra portion of food each day. After two weeks the length of each guppy is again measured and recorded. The guppies that received an extra portion of food averaged 10% greater length than those in the other group. This is an example of

(A)   an experiment since the researcher imposed a treatment.
(B)   block design.
(C)   an observation since the researcher did not randomly assign the guppies to a side.
(D)   stratified sampling since there are two distinct groups.
(E)   an explanatory variable.

#37   The accompanying box plots are from a study of the effectiveness of meditation as a method to relieve headaches. The top box plot is the number of headaches suffered by a group of volunteers using meditation. The bottom box plot is the number of headaches suffered by a group of volunteers who didn't use meditation.

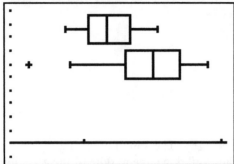

Based on the plots, which of the following statements is TRUE?
I.   The upper quartile range is similar for both groups.
II.   The mean number of headaches in the group that didn't meditate is higher than the mean from the group that did meditate.
III.   A person will have fewer headaches by using meditation.

(A)   I only
(B)   II only
(C)   I and II only
(D)   I and III only
(E)   I, II, and III

#38   Which of the following would increase the power of a statistical test?
I.   Increase the sample size
II.   Reduce the α-level
III.   Shrink the difference between the hypothetical mean and the true mean.

(A)   I only
(B)   II only
(C)   III only
(D)   I and II only
(E)   II and III only

#39   A stock market analyst performed a linear regression on Share Price (independent variable) vs. Total Sales (dependent variable). The value of r^2 was 0.85. Because the analyst wasn't sure that there was a true cause and effect relationship, he switched the variables with Total Sales (independent variable) vs. Share Price (dependent variable). The new value of r^2 is

(A) $\dfrac{-1}{0.85}$

(B)   -0.85

(C)   0.85

(D) $\dfrac{1}{\sqrt{0.85}}$

(E) $\dfrac{1}{0.85}$

#40   In order to conserve water in a dry state, residents are asked to measure how much water their sprinkler system discharges in 15 minutes. They are then encouraged to water according to the following schedule.

Water depth in inches in 15 minutes	0.125	0.187	0.25	0.313	0.375	0.50	0.625	0.75	1.0
Total weekly watering time (in minutes)									
Spring	52	34	26	20	17	13	10	9	6
Summer	104	69	52	41	35	26	21	17	13
Fall	69	51	39	31	26	19	15	13	10

What is the degrees of freedom for a chi-square test of independence between seasonal watering times and sprinkler system output?

(A)   16
(B)   24
(C)   25
(D)   26
(E)   27

#1

The following computer output comes from weekly salary information for a small company. Non-graduates refer to those without a college degree. Graduates refer to those with a college degree.

GENDER	N	MIN	Q1	MED	Q3	MAX	MEAN	STDEV
Non-graduate	10	350	500	800	900	1150	700	125
Graduate	15	400	600	1050	1300	2000	1000	250

(A)   Draw side-by-side box plots of the data. Comment on any similarities or differences between the plots.

(B)   Suppose that a graduate and non-graduate were each selected at random many times and the combined salary of the pair was computed. What would be the mean and standard deviation of the combined salary?

(C)   Suppose that the next 15 non-graduate workers were hired at the median salary for non-graduates. What will be the new mean salary of non-graduate workers?

(D)   If the next five graduate workers are also hired at the median salary of graduates, what will be the new mean salary of the graduate workers?

(E)   The mean salary for each worker before all the new people were hired was $880. With all the new workers hired as indicated in parts (C) and (D), the new mean salary for the company is now $872. Explain how this can happen in light of the results from parts (B) and (D).

#2

A lumber company harvests trees and then plants new seedlings so there will be trees to harvest in the future. It takes about ten years for a tree to mature to cutting size. Each year the company harvests 10% of the trees from that land it owns. Most of the time, the company *clear cuts* an area, meaning they cut down all the trees in a certain plot of ground before replanting the entire parcel. The lumber company believes that this provides the most lumber from a parcel of land. An environmental group believes that selective cutting and replanting actually produces more lumber from a parcel.

(A)   Design a twenty-year study to determine which method will produce the most lumber. You have access to 40 three-acre parcels on a mountainside in Idaho. Half the plots are on the north side of the mountain and half are on the south side. All parcels are essentially similar in all major respects. Be specific in your description. You may use diagrams in your explanation.

(B)   What type of hypothesis test would you use to determine which method was better? Why?

A study of sales data (in billions of dollars) from the sporting goods industry for a 12-year period from 1981 through 1992 (Year 1981 = 1) was investigated and a least squares regression line was computed. A summary of the analysis is shown below.

SALES = 15.5 + 2.68 YEAR
R-SQ = 0.91
N = 12
A residual plot is shown below.

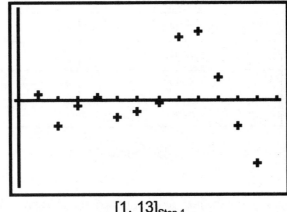

$[1, 13]_{\text{Step 1}}$

(A)   Compute the estimated sales for the year 1986.

(B)   Is the estimated value found in part (A) an over-estimate or an under-estimate of the actual sales. Why?

(C)   The variance of the independent variable (YEAR) is 13.03. What is the variance of the dependent variable (SALES)?

(D)   Based on the residual plot, is the linear model a good model for the entire 12 years? Explain.

(E)   A linear regression t-test was performed on the hypotheses $H_0: \beta = 0$ vs. $H_a: \beta \neq 0$ where β is the slope of the regression line. The value of the t-statistic was 10.03 with 10 degrees of freedom. Is this strong evidence to reject the null hypothesis? Explain.

The table below represents the before and after scores of 424 at-risk students randomly chosen from a large school district that implemented a new strategy to teach reading.

Reading level (grade level)	September 2001	March 2002
Below 5.5	7.0%	4.2%
6	14.9%	5.9%
7	11.1%	9.7%
8	32.3%	27.5%
9	26.9%	30.2%
10+	7.8%	22.5%
Totals	100%	100%

(A)   Explain why the mean reading level cannot be determined.

(B)   The English department claimed that the new reading program had improved the reading ability of the students. At the 5% level of significance, test the claim that the new program has improved reading scores.

#5

The following test scores were obtained randomly from two different, large, introductory college classes in the same subject.

Class 1

4	14	21	22	24	24	26	30	40	40	46	61	65
66	67	70	73	82	85	87	90	94	95	98	98	100

Class 2

18	25	29	30	46	54	55	57	60	61
65	83	84	86	89	89	92	98	100	100

(A)   Draw back-to-back stem and leaf plots of each class's scores. Comment on any similarities and differences between the scores from the two classes.

(B)   If 70% is considered to be a passing grade, test the claim that the proportion of students passing is the same in each class.

One of the responsibilities of the National Collegiate Athletic Association (NCAA) is to verify that member institutions comply with rules regarding the recruitment and awarding of scholarships and benefits to student athletes. Colleges and universities are required to submit annual reports about their activities. To insure that the information is being reported accurately, the NCAA randomly selects several schools each year for an on-campus audit and review of documentation.
Because some universities have been cited for previous non-compliance, they have agreed to be audited more often than usual.

Due to the time and expense required, only 12 universities can be audited each year. The auditors have divided the members into three groups
Group A: Universities on probation for previous non-compliance. Three of these will be audited.
Group B: Universities with large athletic budgets. Four of these will be audited.
Group C: Universities with small athletic budgets. Five of these will be audited.

Let "Yes" mean the university athletic department is in compliance with NCAA guidelines and "No" mean it isn't in compliance. The information is summarized in the following table. The auditors are not aware at the time of their audit if an athletic department is in compliance.

Group A		Group B		Group C			
School #	Compliance	School #	Compliance	School #	Compliance	School #	Compliance
1	Yes	1	No	1	Yes	11	No
2	No	2	Yes	2	No	12	No
3	No	3	No	3	Yes	13	Yes
4	Yes	4	No	4	Yes	14	No
5	Yes	5	Yes	5	No	15	No
		6	Yes	6	No	16	Yes
		7	No	7	No	17	Yes
		8	No	8	Yes	18	Yes
		9	Yes	9	Yes	19	No
		10	No	10	No	20	No

(A)   Use the following random number table to choose a stratified random sample of size 10. Begin on the first line and explain your process. Indicate clearly which numbers you choose.

08421	44753	77377	28744	75592	08563	79140	92454	14592	98481	15373
12109	47806	11421	66812	53045	66831	68098	40772	21558	47781	33586
79177	06928	70708	23822	13048	12633	19687	05233	16853	14374	40011

(B)   Use your sample results to *estimate* the proportion of the entire population of all schools that are in compliance.

(C)   Using the data from the table, how does this estimate compare to the *true* proportion of all schools in compliance?

(D)   Explain why you cannot use the estimate found in part (B) to construct a confidence interval for the true population proportion of schools in compliance.

Solution Keys

#1   Number of programs sold $= (0.40)(800) + (0.50)(1200) = 320 + 600 = 920$

P(Program purchased) $= \dfrac{920}{2000} = 0.46$ ANS: <u>D</u>

#2   Personal Happiness

	Happy	Problems	Total
Died	315	54	369
Lived	8685	1071	9756
Total	9000	1125	10125

Use χ^2 to test relationship

$\chi^2 = 4.81$

$P(\chi^2 \geq 4.81) = 0.028$

Alternately, use a 2 proportion z-test
P(Died|Happy) $= 0.035$, $n_1 = 9000$
P(Died|Not Happy) $= 0.048$, $n_2 = 1125$

$$z = \dfrac{0.035 - 0.048}{\sqrt{(0.036)(0.964)\left(\dfrac{1}{9000} + \dfrac{1}{1125}\right)}} = \pm 2.19$$

$P(z < -2.19 \text{ or } z > 2.19) = 0.028$ ANS: <u>A</u>

#3   In the event of a non-response the correct procedure is to select another respondent using a random number.

         ANS: <u>D</u>

#4 $P(X < 66) = P\left(Z < \dfrac{66 - 68.04}{2.4}\right) = P(Z < -0.85) = 0.1977$

$P(X < 66 \text{ and } X < 66) = 0.1977^2 = 0.039 \approx 0.04$

         ANS: <u>A</u>

#5   The distribution of the sum of 100 spins from a discrete random variable will be a sampling distribution with a mean value of 550. As the number of trials increases, the shape of the sampling distribution becomes approximately normal.

         ANS: <u>A</u>

#6   Choice I   False.

$$P(p_{with} < p_{without}) = P(z < \frac{0.30 - 0.417}{\sqrt{(0.37)(0.63)\left(\frac{1}{40} + \frac{1}{60}\right)}}) = P(z < -1.18)$$

P(z < -1.18) = 0.118

Choice II   True.   Two-proportion z-interval = (-0.31, 0.07)

Choice III   False.   If confidence interval contains zero, there is no statistically significant difference between the two proportions.

Answers I and III are not correct. Corresponds with choice D.

ANS: <u>D</u>

#7   P-value is the probability of the χ^2 statistic = 0.18641, which shows some relationship between the weight of the individual and the weight the individual can lift. However, the relationship is weak.

ANS: <u>D</u>

#8   Choice I   True.
      Choice II   False. $\bar{x}_A = 24.65$, $\bar{x}_B = 26.43$
      Choice III   True.
      I and III only

ANS: <u>D</u>

#9   Choice I   True.   There appears to be some relationship.
      Choice II   True.   No assignment was made.
      Choice III   False.   There is no proof that the relationship was causal.
      Choice IV   False.   No pairing was mentioned.
      Incorrect answers are III and IV.   ANS: <u>E</u>

#10   The selection process does not include all possible groups.

ANS: <u>B</u>

#11   Total favorable values = 225 - 175 = 50 $p = \frac{50}{400} = 0.125$

ANS: <u>A</u>

#12   Accuracy of an estimate from a sample depends on sample size, not population size.

ANS: <u>A</u>

#13 $P(\hat{p} < 0.65) = P\left(Z < \frac{0.65 - 0.71}{\sqrt{\frac{(0.71)(0.29)}{100}}} \right) = P(z < -1.32) = 0.0934$

ANS: <u>B</u>

#14   T-score = 8.7363. Degrees of freedom = 4.
       Critical $t_{(0.0005, 4\ df)} = 8.61$. Use two-tail t-test.
       Therefore, P-value is 0.0005 * 2 = 0.001.

                                                                    ANS: <u>B</u>

#15   December 2002 implies x = 35.
       $\hat{y} = 2.7 + 1.01\ (35) = 38.05$ kw

                                                                    ANS: <u>D</u>

#16   Random numbers chosen are 0 <u>7</u> <u>1</u> 1 9   9 7 <u>3</u> 3 <u>6</u> giving 7,1,3,6, which corresponds to Klein,
       Zavala, Lonnquist, and Elifritz.

                                                                    ANS: <u>C</u>

#17   InvNormal (0.90) = 1.28
       $1.28 = \dfrac{40 - 38}{\sigma} \Rightarrow \sigma = \dfrac{40 - 38}{1.28} = 1.56$ inches

                                                                    ANS: <u>E</u>

#18   Power deals with the ability of a hypothesis test to detect a significant difference between the
       actual mean and the proposed mean. One way this can be achieved is by increasing the sample
       size.

                                                                    ANS: <u>C</u>

#19 $\sigma = \sqrt{Var_1 + Var_2} = \sqrt{4^2 + 7^2} = \sqrt{65}$

                                                                    ANS: <u>D</u>

#20   A simple random sample requires that each potential member be numbered in some identifiable
       way. Each member must have an equal chance of being selected *and* each potential subset must
       have an equal chance of being selected.

                                                                    ANS: <u>E</u>

#21   This is a geometric distribution.
       $P(X = 4) = (0.7)^3(0.3) = 0.1029$

                                                                    ANS: <u>B</u>

#22   Increasing the sample size increases the power of a hypothesis test. Decreasing the sample size
       decreases the power of a hypothesis test.

                                                                    ANS: <u>B</u>

#23   InvNormal (0.95) = 1.645
       $1.645 = \dfrac{x - 55}{4} \Rightarrow x = 1.645(4) + 55 = 61.58$. Closest answer is 62 wpm.

                                                                    ANS: <u>D</u>

#24   Without a control group, the effectiveness of any experiment cannot be determined. There could be several confounding variables other than the diet that might be responsible for the weight loss.

ANS: D

#25   P-value is the probability of randomly obtaining a mean of a sample as extreme or more so as that found in the sample, if the population mean was the stated value.

ANS: C

#26   Linear regression equation is $\hat{y} = 11.8 - 1.6x$.
Residual value $= y_{actual} - y_{predicted}$
Residual of point A = 0.8
Residual of point B = 0
Residual of point C = -1.8
Residual of point D = 0.4
Residual of point E = 0.6
Point C has the largest absolute value.

ANS: C

#27   This is a probability distribution. Let $\bar{x}$ = mean salary.
$\bar{x} = (0.30)(75,000)+(0.50)(54,000)+(0.20)(36,000) = \$56,700$
Mean donation $= (0.05)(\$56,700) = \$2,835$
Total donation $= (\$2,835)(50) = \$141,750$

ANS: E

#28   The question asks for the difference in the number of marks made by each hand. The appropriate test is a one-sample t-test with H_0: $\mu = 0$ vs. H_a: $\mu \neq 0$.
The differences are 3, 13, -19, -1, 12, 12, -4, 11, 14, -14. $\bar{x} = 2.7$, s =11.94.

$t = \dfrac{2.7 - 0}{\dfrac{11.94}{\sqrt{10}}} = \pm 0.715$. $P(t < -0.715 \text{ or } t > 0.715) = 0.493$

ANS: C

#29   Residual plots should be randomly scattered with no apparent pattern. #1, #2, and #4 have a definite pattern. #5 is not scattered. #3 is the one that appears to be randomly scattered.

ANS: C

#30 $C.I = \bar{x} \pm t^{*}_{(0.98,9df)} \dfrac{s}{\sqrt{n}}$

$C.I.= 32.7 \pm 2.821\dfrac{4.3}{\sqrt{10}} = 32.7 \pm 3.84$

ANS: E

#31   Confidence intervals are used to *estimate* the value range of a population mean from a sample. There is no need to compute a confidence interval if the population data is already known.

ANS: C

#32 $P(\text{Suburban}|\ 8\text{days vacation}) = \dfrac{32}{71} = 0.31$

ANS: <u>B</u>

#33   P(Not donuts or Not cinnamon rolls) = 1 - (0.30 + 0.40 + 0.10) = 0.20 or 20%

ANS: <u>B</u>

#34   No treatment is imposed. Workers voluntarily decided how to divide themselves.

ANS: <u>B</u>

#35 Three orchards have yields between 60 and 80 bushels/acre. Two orchards have yields between 80 and 100 bushels / acre. There are a total of 5 of the 20 orchards with yields above 60 bushels /acre = 25%.

ANS: <u>D</u>

#36   The mean weight loss from Looze-Now was actually less than the mean weight loss of the control group. The weight loss must be due to the placebo effect.

ANS: <u>C</u>

#37   A condition of any experiment and the associated hypothesis test is that the data comes from a randomized situation.

ANS: <u>A</u>

#38   The confidence interval for differences must contain only positive values to support a claim of preferential salaries for tall people.

ANS: <u>D</u>

#39   This is not a histogram. It is a time plot with bars. The variable of interest is the number of people employed, which is shown on the vertical axis. Since there are only eight data points, no conclusion can be reach on skewness. The mean is about 5000. So choice III is correct, but no other choices are correct.

ANS: <u>E</u>

#40   Find the critical χ^2 value for $\alpha = 0.05$ and degrees of freedom = 3.

ANS: <u>C</u>

Textbook Correlation for Practice Exam 1 – Multiple Choice

Question	Correct answer	Section	Question	Correct answer	Section
1	D	6.3	21	B	8.2
2	A	13.2	22	B	5.1
3	D	5.1	23	D	2.2
4	A	2.1 and 6.3	24	D	5.2
5	A	9.1	25	C	10.2
6	D	12.2	26	C	3.3
7	D	13.2	27	E	6.3
8	D	1.1	28	C	11.2
9	E	4.2	29	C	4.1
10	B	5.1	30	E	10.1
11	A	6.2	31	C	10.1
12	A	12.2	32	C	6.3
13	B	12.1	33	B	6.3
14	B	14.1	34	B	5.1
15	D	3.3	35	D	1.1
16	C	5.1	36	C	10.4
17	E	2.2	37	A	11.2
18	C	10.4	38	D	11.1
19	D	2.2	39	E	1.1
20	E	5.2	40	C	13.2

#1

(A)

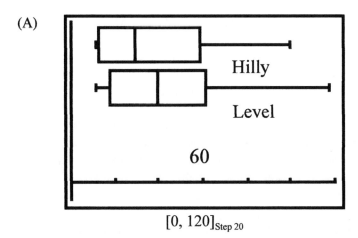

[0, 120]₍Step 20₎

(B)   In viewing graphical displays of data, we want to look for measures of center, shape, and spread. The median for data from hilly ground is 10 (hundred) bushels smaller than the median for data from level ground. However, the means are much closer. Both sets of data are skewed to the right. The InterQuartile Range is similar for both sets of data. Neither data set has an outlier.

(C)   While it appears that level ground produces slightly higher yields, the data are not conclusive. There is a great deal of overlap between the two box plots with similar values for Q1, Median, and Q3.
A t-test for the null hypothesis H_0: $\mu_{Hilly} = \mu_{Level}$ has a P-value of 0.70. This is very weak evidence to support the claim of a difference between the two means.

#2

(A)   The purpose of the study is to determine which campaign is most effective. Since there are two different campaigns and there are also two different marketing situations, one preferred method would be to divide the eight stores east of Main Street and the eight stores west of Main Street into four separate groups of four. Number each of the eight stores east of Main Street from 1-8. Use a random number table to select four stores. These four would get the advertising campaign that stresses convenience. The other four would get the in-store campaign that stresses value. The eight stores west of Main Street would be divided in a similar manner and the campaigns assigned to the two groups. After a reasonable length of time, determine the total sales volume for the new product at each store and compare for similarities or differences.

(B)   There are at least three confounding variables. 1. Stores may serve customers from beyond their neighborhood. If a store is on a busy street, it may have customers with differing incomes than from the surrounding areas. 2. Income may vary by age. There may be age preferences for or against the product. 3. There is no proof that all people would like this product equally. There may be a connection between income and the desire to consume the product in the first place.

(C)   To minimize the effects of confounding variables 2 and 3, a chi-square test for independence would be most useful. There is no convenient way of eliminating possible effects from variable 1, short of asking customers if they live in the neighborhood and deleting any record of purchases from non-neighborhood customers.

#3

(A)   The plot of latitude and average temperature has a stronger relationship. A scatter plot for this graph has a more well-defined trend than the plot using longitude. Additionally, the value of r^2 for latitude is 0.61 which shows a moderate to strong relationship between the explanatory and response variables. The value for r^2 for the longitude graph is 0.0031, which indicates virtually no relationship between variables. Although a residual plot is not shown, a reasonable approximation of the residual plot for latitude can be made by drawing the regression line on the graph. The values above and below the regression line form no apparent pattern. This supports the idea that a line of best fit is a good model to use.

(B) $r^2 = 0.61 \Rightarrow r = + \, or - \sqrt{0.61} = -0.78$. The negative value was chosen since the slope of the regression line is negative.

(C)   Latitude equation is a better predictor. $AvgTemp = 100.5 - 1.83 * 42.3 = 100.5 - 77.4 = 23.09^0$

(D)   Residual = Actual Value - Predicted Value
Predicted Value = $23.5 + 0.0498 * 105.3 = 28.7°$
Actual Value = 15
Residual = $15 - 28.7 = -13.7°$

(E) $P(t \leq -4.495) < 0.0005$.
This is very strong evidence against the null hypothesis. The probability of getting a sample of this size (n = 15) from a population with a true slope of 0 that yields a slope of -1.83 is less than 0.0005, which is less than 1 chance out of 2000. Note: When doing a linear t-test for regression, the degrees of freedom is n - 2.

#4

(A)   Expected Value for Juice $= \dfrac{(RowTotal)(ColumnTotal)}{GrandTotal} = \dfrac{(143)(288)}{468} = 88$.

(B) $P(\text{Juice}|\text{Cafeteria}) = \dfrac{30}{33+30+5} = \dfrac{30}{68} = 0.44$.

(C) H_0: There is no relationship between location and type of drink purchased.
       H_a: There is some relationship between location and type of drink purchased.

(D)   Appropriate test statistic is χ^2. $\chi^2 = \sum \dfrac{(Observed - Expected)^2}{Expected}$.
       By Calculator, the value of $\chi^2 = 24.58$.
       P-value < 0.0001.
       Degrees of freedom $= (\text{row} - 1)(\text{col} - 1) = (3 - 1)(3 - 1) = 4$.

(E)   Since the P-value is so small, there is very strong evidence that there is some relationship between a person's location and the type of drink purchased.

#5

(A)   Let X = number of shots made out of 12 attempts.
        $P(X = 8) = _{12}C_8 (0.8)^8(0.2)^4 = 495(0.8)^8(0.2)^4 = 0.133$

(B) $\mu = np = 12(0.8) = 9.6$
        Standard deviation = $\sqrt{n(p)(1-p)} = \sqrt{12(0.8)(0.2)} = 1.39$.

        Using the Empirical Rule, approximately 95% of all values lie within 2 standard deviations of the mean. The distribution of made shots out of twelve attempts is a binomial distribution that is symmetric in shape and centered at the mean of 9.6. For large values of n the distribution is approximately normal in shape. The 95% range is the middle 95% and is found using a confidence interval.
        C.I. = $9.6 \pm 2(1.39) = 9.6 \pm 2.78 = 6.82 \le X \le 12.38$. This is rounded to 7 to 12.

(C)   No. In 13.3% of all similar trials of 12 shots, Corrinna would make exactly 8 shots. This is a relatively large percentage, occurring about one out of every seven similar situations. It does not appear that nervousness made her miss 4 of 12 shots.

(D) $P(X = 60) = _{90}C_{60} (0.8)^{60}(0.2)^{30} = 0.001$
        If Corrinna's proportion of success were 0.8, she could expect to make exactly 60 of the next 90 shots about 1 time in a 1000 trials of 12 shots. It appears her probability of success is not as high as 0.8.

(A)   Each section has a unique number assigned to it which allows the use of a random number. It is not a requirement of a Simple Random Sample that numbers be assigned in a particular order, only that each entry has a unique number assigned. Random numbers from 1-36 can now be assigned to select a section as needed.

(B)   Total Expected Revenue = 35($1,000) + 1($100,000) = $135,000.

Expected value for each randomly selected section = $\dfrac{\$135,000}{36}$ = $3,750.

(C)   Trial #1. Start with row 1. Use two digit numbers.

29676	20591	68086	26432	46901	20849	89768	81536	86645	12659
92259	57102	80428	25280	00742	57392	39064	66432	84673	40027
32832	61362	98947	96067	64760	64584	96096	98253	05366	04213

Sections selected are 29, 5, 8, 32, and 12. Total revenue received = $104,000.

Trial #2. Start with row 2. Use two digit numbers.

29676	20591	68086	26432	46901	20849	89768	81536	86645	12659
92259	57102	80428	25280	00742	57392	39064	66432	84673	40027
32832	61362	98947	96067	64760	64584	96096	98253	05366	04213

Sections selected are 25, 2, 6, 32, 34 (Second occurrence of 25 is ignored).
Total revenue received = $5,000.

(D)   The mean of the distribution is the Expected Value for each section multiplied by the number of sections chosen in each sample. μ = ($3,750)(5) = $18,750.

#1   Left whisker is much longer than right whisker, which indicates skewness to the left.

ANS: <u>E</u>

#2   The station wants to predict the proportion of "registered voters."

ANS: <u>B</u>

#3   The reporter averaged the two proportions. No weighting was given to the extreme differences in populations.

ANS: <u>A</u>

#4 $n = \left[\dfrac{z*_{0.95}\, \sigma}{ME} \right]^2 = \left[\dfrac{1.96\sqrt{(0.88)(0.12)}}{0.04} \right]^2 = (15..93)^2 = 253.55 \approx 254$

ANS: <u>B</u>

#5 $p_S = \dfrac{68}{128} = 0.53$

$p_E = \dfrac{45}{110} = 0.41$

$\hat{p} = \dfrac{68 + 45}{128 + 110} = 0.475$

$z = \dfrac{0.53 - 0.45}{\sqrt{(0.475)(0.525)\left(\dfrac{1}{128} + \dfrac{1}{110}\right)}} = \pm 1.88$

$P(z < -1.88 \text{ or } z > 1.88) = 0.0599 \approx 0.06$

ANS: <u>B</u>

#6   Mean is approximately 4 with a range from 0 to 8. Data are approximately normal in shape. This eliminates A, D, and E. Value of histogram for $X = 0$ homes is a larger number than the value of histogram for $X = 8$, so point for $X = 0$ must be above the line. This leaves choice B.

ANS: <u>B</u>

#7 $(10, 29)$ appears to be on the same trend line as the rest of the data. If $(10, 29)$ is removed, the linear regression line does not change appreciably, so $(10, 29)$ is NOT an influential observation.

ANS: <u>A</u>

#8   Matched pairs should keep all differences due to gymnasts as small as possible which will reduce the likelihood of a confounding variable.

ANS: <u>D</u>

#9   The mean of a sampling distribution is equal to the mean of the population. The shape of a sampling distribution is always approximately normal with a standard deviation of $\frac{s}{\sqrt{n}}$.
However, the standard deviation is not stated in this example.

ANS: E

#10   The data are strongly concave upward. To linearize such data, it is necessary to apply a function that is increasing and concave downward. A square root function or the log of the dependent variable will accomplish this. I and III are appropriate.

ANS: B

#11   The standard deviation of the difference (or sum) between two means is
$$\sqrt{Var_1 + Var_2} = \sqrt{2^2 + 3^2} = \sqrt{13}$$

ANS: C

#12   A confidence interval uses a sample mean to estimate a range for a population mean. The writer already has the information about the population. No confidence interval is necessary.

ANS: B

#13   Width of interval is (3 - 1.75) = 1.25. Height is 0.5. Total area that is equal to the probability is (1.25)(0.5) = 0.625.

ANS: C

#14   Using the most conservative degrees of freedom, $C.I.= (127 - 134) \pm 2.12\sqrt{\dfrac{23^2}{20} + \dfrac{20^2}{17}} = -7 \pm 14.98$
By calculator the confidence interval is -7 ± 14.35.

ANS: E

#15   Since the confidence interval did not contain 0, the difference between the two groups is statistically significant. Since all the values in the confidence interval are positive, then the college player group has a higher mean than the non-player group.

ANS : B

#16   This is a Cumulative Geometric Distribution.
P (X < 5) = (P(X = 4) + P(X = 3) + P(X = 2) + P(X =1)
         = 0.0864 + 0.144 + 0.24 + 0.4 = 0.8704

ANS: D

#17　　$\hat{p} = \dfrac{891}{6000} = 0.1485$

　　　　$z = \dfrac{0.1485 - 0.13}{\sqrt{\dfrac{(0.13)(0.87)}{6000}}} = 4.26$

　　　　$P(z > 4.26) < 0.0001$

ANS: <u>A</u>

#18　　InvNormal(0.10) = -1.28 = z

　　　　$-1.28 = \dfrac{x - 139}{26} \Rightarrow x = -1.28(26) + 139 = 105.72 \approx 106$

ANS: <u>B</u>

#19　　Treatment is the storage temperature. Experimental unit is a pill. Time to relief from headache pain is the response variable.

ANS: <u>E</u>

#20　　$\hat{p}_1 = \dfrac{12}{75} = 0.16$

　　　　$\hat{p}_2 = \dfrac{9}{45} = 0.20$

　　　　$C.I = (0.16 - 0.20) \pm 1.684\sqrt{\dfrac{(0.16)(0.84)}{75} + \dfrac{(0.20)(0.80)}{45}} = -0.04 \pm 0.12$

ANS: <u>D</u>

#21　　Since a constant was added to each value, the spread and standard deviation will remain the same.

ANS: <u>B</u>

#22　　$P_{(DVD)} = (0.70)(0.40) + (.30)(0.30) = 0.28 + 0.09 = 0.37$

ANS: <u>D</u>

#23　　$\chi^2 = \dfrac{(Observed - Expected)^2}{Expected}$, $Expected = \dfrac{(348)(117)}{770} = 52.88$

　　　　$\chi^2 = \dfrac{(52 - 52.88)^2}{52.88} = 0.015$

ANS: <u>B</u>

#24　　Data are trending upwards, but the data are not tightly grouped, which indicates a weak, positive correlation.

ANS: <u>A</u>

#25   InvNormal (0.95) = 1.645 = z

$$1.645 = \frac{x - 750{,}000}{150{,}000} \Rightarrow x = 150{,}000(1.645) + 750{,}000 = 996{,}750$$

ANS: <u>B</u>

#26   InterQuartile Range = Q3 - Q1 = 107 - 88 = 19.
Upper Outlier = Q3 + 1.5*IQR = 107 +1.5(19) = 135.5
The smallest integer value greater than 135.5 is 136.

ANS: <u>D</u>

#27   Middle 95% refers to a 95% confidence interval. Using the Rule of Thumb, the middle 95% lies within two standard deviations of the mean. The standard deviation for this sample is $\frac{10}{\sqrt{25}} = 2$.
Therefore, the 95% confidence interval should range 4 seconds on either side of the sample mean or 44 to 52 seconds.

ANS: <u>D</u>

#28   Choice I   True
Choice II   False. $Max_{City\ B}$ is much greater than the $Max_{City\ A}$
Choice III   True. IQR = 61.5 - 9.5 = 52. Q3 + 1.5*IQR = 61.5 + 78 = 139.5.
I and III

ANS: <u>E</u>

#29 $C.I. = 8.7 \pm 2.262 \left(\frac{3.2}{\sqrt{10}} \right) = 8.7 \pm 2.29$

ANS: <u>C</u>

#30   A stratified sample randomly selects values from previously defined subgroups.

ANS: <u>E</u>

#31   There are 11 differences to be used for the hypothesis test. Degrees of freedom = n - 1 = 11 - 1 = 10.

ANS: <u>A</u>

#32   Power is a measure of a hypothesis test to detect a difference between the actual mean and the hypothesized mean. Increasing the sample size increases the power. Larger differences require smaller sample sizes to be detected.

ANS: <u>B</u>

#33 $Expected = \dfrac{(RowTotal)(ColTotal)}{GrandTotal} = \dfrac{(63)(81)}{150} = 34.02 \approx 34$

In this type of computer output the expected value is directly beneath the observed value in each cell. The correct answer is 34.

ANS: <u>B</u>

#34 If (20, 21) is removed from the data set, both $\bar{x}$ and $\bar{y}$ will decrease. The slope will also decrease, which will cause the y-intercept to rise since the regression line always passes through $\left(\bar{x}, \bar{y}\right)$.

ANS: <u>C</u>

#35 Using Venn Diagrams

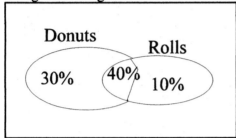

P(Donuts or Rolls but not both) = 30% +10% = 40%

ANS: <u>C</u>

#36 Choice I   False
Choice II   False
Choice III   True
III only

ANS: <u>C</u>

#37 Difference will be about -5 with a normal distribution. Graph C fits this the best.

ANS: <u>C</u>

#38 Degrees of freedom = (rows - 1)(columns - 1) = (4 - 1)(5 - 1) = 12

ANS: <u>C</u>

#39 The purpose of blocking is to eliminate possible confounding variables. There might be a specific genetic difference between Pacific and Atlantic salmon that causes a difference in results. Although we don't know for sure that such a condition might occur, we have to plan for its possibility.

ANS: <u>B</u>

#40 $\hat{p}_{Airport} = \dfrac{23}{65} = 0.354$, $\hat{p}_{Concert} = \dfrac{18}{52} = 0.346$

$$z = \dfrac{0.354 - 0.346}{\sqrt{(0.35)(0.65)\left(\dfrac{1}{65} + \dfrac{1}{52}\right)}} = \pm 0.09$$

$P(\, z < -0.09 \text{ or } z > 0.09) = 0.931$

ANS: E

Textbook Correlation for Practice Exam 2 – Multiple Choice

Question	Correct answer	Section	Question	Correct answer	Section
1	E	1.1	21	B	1.2
2	B	5.1	22	D	6.3
3	A	12.2	23	B	13.2
4	B	12.1	24	A	3.2
5	B	12.2	25	B	2.2
6	B	1.1	26	D	1.2
7	A	3.3	27	D	10.1
8	D	5.2	28	E	1.2
9	E	9.1	29	C	11.1
10	B	4.1	30	E	5.1
11	C	11.2	31	A	10.4
12	B	10.1	32	B	10.4
13	C	6.1	33	B	13.2
14	E	11.2	34	C	3.3
15	B	11.2	35	C	6.3
16	D	8.2	36	C	10.4
17	A	12.1	37	C	9.3
18	B	2.2	38	C	13.2
19	E	5.2	39	D	5.2
20	D	12.2	40	E	12.2

Answer Key for Practice Exam 2
Free Response

#1

(A)   There are three averages: mean, median, and mode. Each is a measure of center. The median is
the middle value in a sorted list. People do not come in fractional parts. The answer must be an
integer unless there is an even number of subjects. Then the median is the mean of the two
adjacent integer values in the middle of the list. In this case the mean of the two numbers would
either be an integer or would have to end in .5. The mode is the most common value. Again, it
would have to be an integer. Only the mean could have decimal values in this setting.

(B)   Number of people actually attending work = (200)(0.91) = 182.

Number of cars needed to accommodate 182 people = $\frac{182}{2.3}$ = 79.1 ≈ 80 *cars*

(C) $P(miles > 22) = P(z > \frac{22 - 19.6}{6.3}) = P(z > 0.38) = 0.352$

P(2 cars travel more than 22 miles) = P(miles > 22) and P(miles > 22) =
(0.352)(0.352) = 0.124.

#2

(A)   Number the pigeons from 1 - 60. Randomly select 30 pigeons for one group and 30 for another group. On a sunny day, release the 30 pigeons from a location that is beyond visual range from the coop. On a cloudy day release the other 30 pigeons from the same location. Tally how many pigeons returned to the coop within a day. This plan assumes that not all pigeons may return to the coop.

An alternate method is to time how long it takes for each pigeon to return to the coop. Each of these pigeons would have to be numbered and released at specified intervals. This would be done for the first group on a sunny day. The second group would be released on a cloudy day. This plan assumes that all pigeons will return.

(B)   If the first plan is used, the appropriate test is a two proportion z-test on the proportion of each group that returned to the coop.
Assumptions are:
1.   Each group is taken from a random sample.
2.   Each pigeon's behavior is independent of other pigeons. Each pigeon navigates its own way back to the coop and doesn't depend on other pigeons.
3. $n_s \hat{p}_s \geq 5$ and $n_s (1 - \hat{p}_s) \geq 5$
      $n_c \hat{p}_c \geq 5$ and $n_c (1 - \hat{p}_c) \geq 5$

The second plan would use a two-sample t-test on the mean time required to return to the coop.
Assumptions are
1.   Each group is taken from a random sample.
2.   Each pigeon's behavior is independent of other pigeons. Each pigeon navigates its own way back to the coop and doesn't depend on other pigeons.
3.   The times for both data sets should be reasonably normal in shape. This can be check by drawing a dot plot, stem and leaf plot, box plot, or histogram.

#3

(A)   No. If they had estimated the actual number of hours of sleep the slope would have been 1. In this case the slope was 0.907 and the y-intercept was 0.03 which means they were predicting about 91% of the actual number of hours of sleep they actually received. The volunteers were consistently under-estimating the number of hours of sleep actually obtained.

(B)   The coefficient of determination is r^2, which is equal to 0.938. That means that about 94% of the variation of the actual hours of sleep from the mean of the actual hours of sleep can be explained by the least squares regression of Actual Hours on Estimated Hours. In other words, as the Estimated Hours increases, the equation explains (or predicts) about 94% of the change in the Actual Hours of sleep.

(C)   Residual = Actual y-value - Predicted y-value
        Predicted y-value is 0.032 + 0.907(8.2) = 7.47
        Residual = 8.2 - 7.47 = 0.73 hours

(D)   A least squares regression always passes through the point $(\bar{x}, \bar{y})$.

$$\bar{y} = 0.032 + 0.907\bar{x} \Rightarrow \bar{x} = \frac{\bar{y} - 0.032}{0.907} = \frac{6.70 - 0.032}{0.907} = 7.35 hrs$$

(E)   From the scatter plot it is apparent that the data strongly follow a linear pattern with no extreme outliers or influential observations. The correlation coefficient is very close to 1, which is another indicator of a very strong linear fit. The residual plot will have points randomly scattered on either side of the x-axis. They will be relatively close to the x-axis. The most extreme residual has a value of -0.66. Also, there is no convergence or divergence of the residuals as the x-value increases. Which means the line is a good fit for the entire set of data.

(F)   The correlation coefficient is a measure of the strength of the relationship between two variables. If the position of the variables is switched, the strength stays the same. Therefore, the value of the correlation coefficient, r, is $\sqrt{0.938} = 0.969$.

#4

(A)   The question asked for the difference for *each* worker. This will involve finding the difference for each worker (New method - old method).

Worker #	1	2	3	4	5	6	7	8	9	10
Difference	-1	2	1	4	-4	-5	-3	3	5	5
Worker #	11	12	13	14	15	16	17	18	19	20
Difference	12	-1	3	3	-2	7	2	4	1	3

$$\bar{x}_d = 1.95, s_d = 3.98$$

$$C.I. = \bar{x}_d \pm t^*_{(0.95,19df)} \frac{s_d}{\sqrt{n}} = 1.95 \pm 2.093 \left(\frac{3.98}{\sqrt{20}}\right) = 1.95 \pm 1.86 \Rightarrow 0.09 \le u_d \le 3.81$$

(B)   An increase of 3 figurines per worker per day is not considered significant since the value falls within the 95% confidence interval.

(C)   Claim:

The mean population increase in the number of figurines per worker is greater than 0.

Counterclaim:

The mean population increase in the number of figurines per worker is not greater than 0.

Hypotheses:

$H_0: \mu_d > 0$ vs. $H_a: \mu_d = 0$

Where μ_d represents the mean difference of new method - old method.

Level of significance :

Let $\alpha = 0.05$ (This is an arbitrary value).

Conditions and assumptions:

(A)   Data are from an SRS. True, stated in problem.

2.   Data are independent. True, it is unlikely that the number of figurines painted by one person will affect the number of figurines painted by another person.

3.   The sample data are not non-normal. A dot plot shows a peak at about 3 with a wide spread from -5 to 7 with a single outlier at 12. There are no extreme departures from normality. A stem and leaf plot or some other quick display of data could also be used.

Conditions and assumptions are verified.

Use a right-tailed, one-sample t-test with 19 degrees of freedom.

$$t = \frac{\bar{x} - \mu}{\frac{s}{\sqrt{n}}} = \frac{1.95 - 0}{\frac{3.98}{\sqrt{20}}} = 2.19 \quad , \quad P(t \ge 2.19) = 0.02$$

Decision:

Since P-value is less than stated α value, reject H_0.

Conclusion:

There is sufficient evidence to support the claim that there was an increase in the mean number of figurines painted by each worker. It appears the new method increases the productivity of the workers.

#5

(A) Use a tree diagram to determine possible cases.

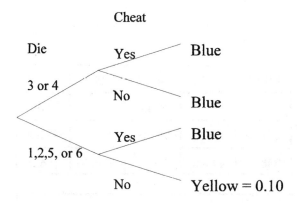

$P(3 \text{ or } 4) = \frac{1}{3}$. $P(1,2,5, \text{ or } 6) = \frac{2}{3}$.

$P(\text{Not cheat}) = X$. So $\frac{2}{3}X = 0.10 \Rightarrow X = 0.15$

$P(\text{Cheat}) = 1 - 0.15 = \underline{0.85}$

(B) $C.I = \hat{p} \pm z^{*}{}_{0.90}\sigma_{\hat{p}} = 0.85 \pm 1.96\sqrt{\dfrac{(0.85)(0.15)}{45}} = 0.85 \pm 0.104 \Rightarrow 0.746 \le p \le 0.954$

(A) One method is to assign the following numbers to the given blood type.

        A: 01-40   B: 41-50   AB: 51-55   O: 56-00

Starting with Row 1, mark off two-digit numbers and assign them to a specific blood type group. At the same time, assign the following numbers to the Rh factor,

        Rh Positive: 1-85   Rh Negative: 86-00

Beginning with Row 2, mark off two-digit numbers and assign them to a specific Rh factor for the corresponding blood type selection.

Row 1	10480	15011	01536	02011	81647	91646	69179	14194	62590	36207
Row 2	20969	99570	91291	90700	22368	46753	25595	8533	30995	89198
Row 3	27982	53402	93965	34095	52666	19174	39615	99505	24130	48360
Row 4	22527	97265	76393	64809	15179	24830	49340	32081	30680	19655

First random #	Blood type	Second random #	Rh factor	First random #	Blood type	Second random #	Rh factor
10	A	20	Pos	79	O	22	Pos
48	B	96	Neg	16	A	36	Pos
01	A	99	Neg	46	B	84	Pos
50	B	95	Neg	69	O	67	Pos
11	A	70	Pos	17	A	53	Pos
53	AB	91	Neg	91	O	25	Pos
60	O	29	Pos	41	B	59	Pos
20	A	19	Pos	94	O	58	Pos
81	O	07	Pos	62	O	53	Pos
64	O	00	Neg	59	O	33	Pos

Tally of Results: A Positive = 5   A Negative = 1
                         B Positive = 2   B Negative = 2
                         AB Positive = 0   AB Negative = 1
                         O Positive = 8   O Negative = 1

This simulation can be done many other ways. For example, numbers could be read in four-digit groups. The first two determine the blood type and the second two determine the presence or absence of the Rh factor. Also, the assignment of the random numbers can be done in any way that preserves the proportions.

(B)   Use $\hat{p}$ for the best estimate of p.

$$ME = z^*_{0.96} \sqrt{\frac{\hat{p}(1-\hat{p})}{n}} \Rightarrow n = \left(\frac{z^*_{0.96} \sqrt{\hat{p}(1-\hat{p})}}{ME} \right)^2 =$$

$$\left(\frac{2.054\sqrt{(0.45)(0.55)}}{0.03} \right)^2 = 34.06^2 = 1160.08 \approx 1161$$

(C) $$C.I. = \hat{p} \pm z^*_{0.96} \sqrt{\frac{\hat{p}(1-\hat{p})}{n}} = 0.45 \pm 2.054 \sqrt{\frac{(0.45)(0.55)}{20}} \Rightarrow 0.22 \leq p \leq 0.68$$

(D)   A person with A Negative blood can receive a donation from a person with A negative or O Negative. The corresponding proportions for those groups are
A Negative = (0.40)(0.15) = 0.06 and O Negative = (0.45)(0.15) = 0.0675.
P(A Negative or O Negative) = 0.06 + 0.0675 = 0.1275.

#1   22 <u>368</u> 46573 <u>25</u>595 85393 <u>30</u>995 89198 <u>27</u>982 53401
Numbers selected are 22, 36, 25, 30, 27.

ANS: <u>C</u>

#2   Sampling distributions are normally distributed, regardless of the underlying distribution they are drawn from.

ANS: <u>D</u>

#3 $\hat{p}_A = \dfrac{35}{125} = 0.28$, $\hat{p}_B = \dfrac{17}{68} = 0.25$. Since the combined sample size is very large, use $Z^*_{Critical}$ instead of $t^*_{Critical}$.

$$C.I. = (0.28 - 0.25) \pm 1.96\sqrt{\dfrac{(0.28)(0.72)}{125} + \dfrac{(0.25)(0.75)}{68}} = 0.03 \pm 0.13$$

ANS: <u>E</u>

#4   Degrees of freedom = 9-1 = 8. $t^*_{(0.95, 8\ df)} = 2.306$

$$n = \left[\dfrac{t^* \sigma}{ME}\right]^2 = \left[\dfrac{2.306(2.7)}{0.5}\right]^2 = (12.474)^2 = 155.60 \ rounded\ to\ 156$$

ANS: <u>D</u>

#5 $$Expected_{(Fish,\ Hockey)} = \dfrac{(RowTotal)(ColTotal)}{GrandTotal} = \dfrac{(38)(27)}{333} = 3.08$$

ANS: <u>E</u>

#6   Each typist serves as his/her own control group, which minimizes any difference between typists' natural coordination.

ANS: <u>E</u>

#7   Geometric Distribution. Let X be the number of people interviewed until a Democrat is encountered.
$P(X \le 3) = P(X = 1) + P(X = 2) + P(X = 3)$
$\qquad\qquad = 0.60 + (0.40)(0.60) + (0.40)(0.40)(0.60)$
$\qquad\qquad = 0.60 + 0.24 + 0.06$
$\qquad\qquad = 0.936$

ANS: <u>A</u>

#8   Both plots A and E show improvement. A appears to be more normal in its distribution.

ANS: <u>A</u>

#9   Choice I   True
      Choice II   False
      Choice III   False
      I only

ANS: <u>A</u>

#10 Residual plots from a least squares linear regression should be randomly scattered with no noticeable pattern. B, C, and E appear to have a pattern. A shows many points below the axis and just a few points above. D fits the requirements the best.

ANS: <u>D</u>

#11 Because a treatment was imposed, a control group should be set up to guard against any possible confounding variable.

ANS: <u>E</u>

#12 Confidence intervals indicate the reliability of the sampling method.

ANS: <u>D</u>

#13 $\hat{p}_1 = \dfrac{206}{579} = 0.356$, $\hat{p}_2 = \dfrac{28}{132} = 0.212$, $\hat{p} = \dfrac{206 + 28}{579 + 128} = 0.329$

$$z = \dfrac{0.356 - 0.212}{\sqrt{(0.329)(0.671)\left(\dfrac{1}{579} + \dfrac{1}{128}\right)}} = \pm 3.17$$

$P(z < -3.17 \text{ or } z > 3.17) = 0.0015 \approx 0.002$

ANS: <u>A</u>

#14 $n\hat{p} = 96$, $n(1 - \hat{p}) = 4$. The sample size is 100, not 2500.

ANS: <u>C</u>

#15 Width = (-0.60 - -1) = 0.40. Height = 0.5. Probability = (0.40)(0.5) = 0.20.
B is incorrect since probability cannot be negative.

ANS: <u>C</u>

#16 $\sigma = \sqrt{Var_1 + Var_2} = \sqrt{3^2 + 4^2} = 5$

ANS: <u>B</u>

#17 The correlation measures the strength of the linear relationship between the independent and dependent variables. The strength of the relationship doesn't change when the variables switch position.

ANS: <u>D</u>

#18 A multi-stage sample takes an SRS of each full subgroup of a population and then takes an SRS of the remaining subgroups, etc. A stratified sample takes an SRS of several previously defined subgroups and does not go further.

ANS: <u>A</u>

#19 The spokesperson is basing the claim on a 20% proportion. The null hypothesis always contains 0. The alternate in this case needs to be greater than 0.20.

ANS: <u>B</u>

#20 P(Bachelor's degree| income is between \$60,001 and \$80,000) = $\frac{41}{111}$ = 0.37

ANS: <u>D</u>

#21 Side effects = (60)(0.40) + (80)(0.30) = 24 + 24 = 48.

ANS: <u>B</u>

#22 Q3 of the older mice is about the same as Q1 of the younger mice.

ANS: <u>E</u>

#23 A confidence interval is used to *estimate* the value range of the population mean from a sample when the data from the population is unknown. The chief has all the data from the population.

ANS: <u>D</u>

#24 Power is a measure of the ability of a hypothesis test to detect a difference between the actual mean and a hypothesized mean. It can be increased by increasing the sample size.

ANS: <u>D</u>

#25 $\hat{p}_{SLC} = \frac{1800}{1,500,000} = 0.0012$, $\hat{p}_{Chicago} = \frac{1900}{6,000,000} = 0.0003$,

$\hat{p} = \frac{1800 + 1900}{1,500,000 + 6,000,000} = 0.00049$

$z = \dfrac{0.0012 - 0.0003}{\sqrt{(0.00049)(0.99951)\left(\dfrac{1}{1,500,000} + \dfrac{1}{6,000,000}\right)}} = \pm 43.58$

P(z < -43.58 or z > 43.58) ≈ 0

ANS: <u>A</u>

#26 InvNormal(0.75) = 0.67 = z

$z = \frac{x - \mu}{\sigma} \Rightarrow 0.67 = \frac{30 - 22}{\sigma} \Rightarrow \sigma = \frac{30 - 22}{0.67} = 11.9$

ANS: <u>D</u>

#27 Ethnic background is the explanatory variable. Length of time until hired is the response variable. Area of specialty is a possible confounding variable.

ANS: <u>C</u>

#28 Matched pairs helps to eliminate any lurking variables due to any unique characteristics about the subject.

ANS: <u>B</u>

#29 InvNormal(0.99) = 2.326 = z

$2.326 = \frac{120 - \mu}{24} \Rightarrow \mu = 120 - 2.326(24) = 64.18 \approx 64$

ANS: <u>C</u>

#30 $Slope = r\dfrac{s_y}{s_x} = 0.9 * \dfrac{24}{12} = 1.8$. Equation: $\hat{y} - \bar{y} = m(x - \bar{x})$

          $\hat{y} = 1.8(x - 105) + 140 = -49 + 1.8x$

          $\hat{y} = -49 + 1.8(110) = 149$

ANS: <u>D</u>

#31   Dr. Smith = (40% of patients)(40% surgery) = 16%
          Dr. Wilson = (60% of patients)(30% surgery) = 18%
          Total = 34%

ANS: <u>B</u>

#32   The data are increasing and concave downward. Any transformation needs to make the y values larger. Of the choices offered, only squaring the y values will accomplish this.

ANS: <u>B</u>

#33   P(*Post* or *Gazette*) = 0.20 + 0.25 + 0.15 = 0.60
          P(*Tribune*) = 1 - 0.60 = 0.40

ANS: <u>D</u>

#34 $\hat{p} = \dfrac{36}{150} = 0.24$, p = 0.21

          $z = \dfrac{0.24 - 0.21}{\sqrt{\dfrac{(0.21)(0.79)}{150}}} = 0.90$

          P(z > 0.90) = 0.184

ANS: <u>B</u>

#35   Increasing the confidence level will widen the confidence interval and will lessen the chance of not capturing the true mean.

ANS: <u>A</u>

#36   Multiplying all values in a two-way table by a constant causes the value of the original χ^2 to be multiplied by the same constant.

ANS: <u>D</u>

#37   The values are normally distributed with a median near 6. 50% of the values should be below 6 and 50% should be above 6. The values range from about 3 to about 10. The domain of the new graph should have the same range of values. D is the best match based on these guidelines.

ANS: <u>D</u>

#38   The slope = 1.5 mm/kg. Multiplying this by 4 kg gives an increase of 6 mm.

ANS: <u>C</u>

#39   A   True
          B   False.   The median will shift down one player.
          C   True
          D   Mode is still one, but a mode of one is generally ignored.
          E   B is already false.

ANS: <u>B</u>

#40   The equation is HOMEVALU = 35.700 + 0.02875*SQFEET.
        HOMEVALU = 35.700 + 0.02875*2000 = 93.20 (in thousands) = \$93,200.

ANS: <u>A</u>

Textbook Correlation for Practice Exam 3 – Multiple Choice

Question	Correct answer	Section	Question	Correct answer	Section
1	C	5.1	21	B	6.3
2	D	9.3	22	E	1.1
3	E	12.2	23	D	10.1
4	D	10.1	24	D	10.4
5	E	13.2	25	A	12.2
6	E	5.2	26	D	2.2
7	A	8.2	27	C	5.2
8	A	1.1 & 2.1	28	B	11.2
9	A	1.1	29	C	2.2
10	D	4.1	30	D	3.3
11	E	5.2	31	B	6.3
12	D	10.1	32	B	4.1
13	A	12.2	33	D	6.3
14	C	12.1	34	B	12.1
15	C	6.1	35	A	11.1
16	B	10.4	36	D	13.2
17	D	3.2	37	D	11.2
18	A	5.1	38	C	3.3
19	B	10.2	39	B	1.2
20	D	6.3	40	A	3.3

#1

(A)

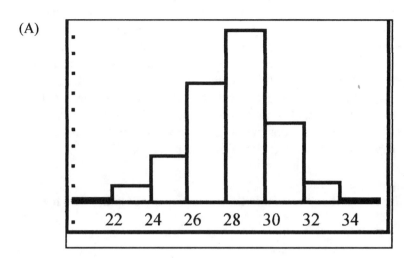

(B) Minimum =   21
     Q1 =   27
     Median =   28
     Q3 =   29
     Maximum =   35

(C) The median is at 28 mph while the mean is 27.89. The graph shows a reasonably symmetric shape with no outliers. The InterQuartile Range is (29 - 27) = 2. This shows a very strong concentration of data near the center with little spread.

(D) InvNormal (85th precentile) = 1.04 = z score
$$z = \frac{x - \mu}{s} = \frac{x - 27.89}{2.36} \Rightarrow x = 27.89 + (1.04)(2.36) = 30.34 \approx 30\,mph\,.$$

#2

(A)   This design uses a matched pairs design.
Use the four-passenger cars first.
Randomly assign one driver to each car. Flip a coin to determine if the car gets gasoline without the additive first. Otherwise, the car gets gasoline with the additive first. Have a driver drive the car with a full tank of the gasoline chosen by the coin flip. Drive the car over a specific course at a specific speed until the tank is empty. Then refill the tank with the same type of gasoline. Drive the car over the course at a specific speed. Drive until the car runs out of gas. Determine the total miles driven. Divide the miles by the capacity of the tank to determine the mileage. Then fill the tank with the other type of gasoline and drive until the tank is empty to clean out any traces of the previous gas. Then refill the tank with the other type of gas. Drive the car over the course at a specific speed. Drive until the car runs out of gas. Determine the total miles driven. Divide the miles by the capacity of the tank to determine the mileage.

Repeat the process with the six-passenger cars.
Compare the results and see if there is an improvement in the mileage using gasoline with the new additive.

(B)   First design.
Use a two-sample t-test on the mean mileage of the four-passenger car with and without the additive.
Second design.
Use a two-sample t-test on the mean mileage of the six-passenger car with and without the additive.

In both cases the hypotheses are the same.
$H_0: \mu_A = \mu_N$ vs. $H_a: \mu_A > \mu_N$
where μ_A = mean mileage using gasoline with the additive
and μ_N = mean mileage using gasoline without the additive

#3

(A)   From the scatter plot of the data, the data seem to follow a generally linear pattern. A measure of goodness of fit is r^2. In this case $r^2 = 0.92$ is close to 1, which usually indicates a good fit. However, it is also important to look at the plot of the residuals. For a good fit, the residuals should be randomly scattered with no apparent pattern. The scatter plot of the residuals meets these criteria except for a couple of points near 1950. Considering all the information together, the proposed model is reasonable.

(B)   Residual = Actual Value - Predicted Value
Actual Value = 78 inches
Predicted Value = 71.8 + (0.223)(48) = 71.8 + 10.7 = 82.5 inches
Residual = 78 - 82.5 = -4.5 inches
A residual value of -4.5 means that the winning jump in 1948 was about 4.5 inches lower than what would have been expected if the height of the winning jump had followed the trend predicted by all winning jumps.

(C)   *LongJump* = 60 + 3.06 * *HighJump*
LongJump = 60 + (3.06)(90) = 60 + 275.4 = 335.4

(D)   This is most likely an associative relationship. It is unlikely that a large value for a winning high jump *caused* a large value for the winning long jump. It is more probable that both values are linked together through the lurking variables of larger size of the athlete and improved training techniques over the years.

#4

(A)   Only births with a girl as the first birth are considered. We only want to look at Girl-Boy and

Girl-Girl. $p_{(SecondBirthBoy|FirstBirthGirl)} = \dfrac{favorable}{possible} = \dfrac{1111}{1111+926} = \dfrac{1111}{2037} = 0.545$

(B)   Claim:

The proportion of boy births is greater than 0.5, given that the first birth is a girl. This is equivalent to stating the population proportion of boy births as a second birth is greater than the population of girl births as a second birth, given that the first birth is a girl.

Counterclaim:

The population proportion of boy births is not greater than 0.5, given the first birth is a girl. Alternately, the population proportion of boy births as a second birth is not greater than the population of girl births as a second birth, given that the first birth is a girl.

Null and Alternate Hypotheses:
H_0: p = 0.5   vs. H_a: p > 0.5
Where p is the population proportion of boy births, given the first birth was a girl.

Level of significance: Let $\alpha = 0.05$

Conditions and Assumptions:
    (A)   Data come from an SRS: Not stated, but will be assumed for this problem in the absence of any information to the contrary.
    (B)   Data are independent. True.
    (C)   Population is at least 10 times sample size. There are more than 20,370 families with exactly 2 children.
    (D)   Shape of distribution. $np = 2037(0.5) = 1018.5 > 10$
$n(1-p) = 2037(1-0.5) = 1018.5 > 10$

All conditions are met.   Use right tail, one proportion z-test

$$z = \dfrac{\hat{p}-p}{\sqrt{\dfrac{p(1-p)}{n}}} = \dfrac{0.545-0.5}{\sqrt{\dfrac{(0.50(0.5)}{2037}}} = 4.1$$

$P(z \geq 4.1) < 0.0001$ (By calculator P-value = 0.00002)
NOTE: If a two proportion z-test had been chosen, the z-score would be 5.80 with a P-value less than 0.000001. Using either method, the decision is the same.

Decision:

> Reject H_0.

Conclusion:

> There is sufficient evidence to support the claim that the proportion of second boy births is greater than 0.5. If a sample of size 2037 were taken from a population where the true proportion was 0.5, the chances of getting a split this extreme (or even more so) are less than 1 out of 10,000.

#5

(A)
Vote in Region 1
P(Region 1 and Candidate A) = (0.35)(0.70) = 0.245
P(Region 1 and Candidate B) = (0.35)(0.20) = 0.07
P(Region 1 and Candidate C) = (0.35)(0.10) = 0.035

Vote in Region 2
P(Region 2 and Candidate A) = (0.50)(0.20) = 0.10
P(Region 2 and Candidate B) = (0.50)(0.60) = 0.30
P(Region 2 and Candidate C) = (0.50)(0.20) = 0.10

Vote in Region 3
P(Region 3 and Candidate A) = (0.15)(0.25) = 0.0375
P(Region 3 and Candidate B) = (0.15)(0.15) = 0.0225
P(Region 3 and Candidate C) = (0.15)(0.60) = 0.09

Totals
Candidate A = 0.245 + 0.10 + 0.0375 = 0.3825
Candidate B = 0.07 + 0.30 + 0.0225 = 0.3925
Candidate C = 0.035 + 0.10 + 0.09 =   0.225

Candidates A and B will be in the runoff election.

(B) $P = \dfrac{favorable}{total} = \dfrac{0.09}{0.225} = 0.40$

#6

(A)
Trial #1
Starting with row 1

20594	69574	13353	67267	39205	07136	93007	67214	91790	01273
67456	42320	60053	59799	63251	82094	64425	92476	20756	16082
99437	40160	38150	81329	85035	58152	96571	24826	41897	43810

Cookie #	1	2	3	4	5	6	Total
Number of chips	3	4	7	3	4	5	26

Trial #2
Starting with row 2

20594	69574	13353	67267	39205	07136	93007	67214	91790	01273
67456	42320	60053	59799	63251	82094	64425	92476	20756	16082
99437	40160	38150	81329	85035	58152	96571	24826	41897	43810

Cookie #	1	2	3	4	5	6	Total
Number of chips	3	7	4	8	6	8	36

(B) $C.I. = \bar{x} \pm t^*_{(0.95, df=29)} \dfrac{s}{\sqrt{n}} = 30.3 \pm 2.045 \dfrac{8.8}{\sqrt{30}} = 30.3 \pm 3.29 \Rightarrow 27.01 \le \mu \le 33.59$

(C)   No. Due to random variation, there are theoretical cases where thousands of chips could be added to a batch and a single cookie could fail to receive three chips. Increasing the number of chips will lessen the probability that a cookie will not have three chips but will not completely eliminate the possibility. The only way to guarantee is to put an infinite number of chips in each batch which is impossible.

(D)   The Central Limit Theorem allows the use of the sample mean as the population mean for large values of n. We want to eliminate all but the lowest 5%. Since $n = 30$, we can use the z-table for our values.

$z = \dfrac{\bar{x} - \mu_{\bar{x}}}{\dfrac{s}{\sqrt{n}}} \Rightarrow -1.645 = \dfrac{\bar{x} - 30.3}{\dfrac{8.8}{\sqrt{30}}} \Rightarrow \bar{x} = 27.66 \approx 28 \, chips$

#1   P(*Post* or *Gazette*) = 0.20 + 0.15 = 0.35 = 35%

ANS: <u>C</u>

#2   Choice I   False. The residuals are not scattered randomly between positive and negative values.
      Choice II   True
      Choice III   False. Half the snow fall gives about 7 inches of water.
      II only

ANS: <u>B</u>

#3   A   True
      B   True
      C   False
      D   True
      E   True. This is a very large sample size. It is likely that the sample standard deviation will be close to the population standard deviation.

ANS: <u>C</u>

#4   Sampling distributions are always normal in shape regardless of the underlying population distribution. The mean of the sampling distribution will equal the mean of the population. Since most of the values of the values are near 7, the mean is near 7.

ANS: <u>C</u>

#5   The mean difference is 9 coins. The question asks if the difference in one draw is 5 or less.

$$z = \frac{5 - 9}{\sqrt{6^2 + 4^2}} = -0.5547$$

P(z < -0.5547) = 0.29

ANS: <u>B</u>

#6   Distance is the explanatory variable. Height is the response variable.

ANS: <u>D</u>

#7   There is no requirement regarding the number of columns and rows.

ANS: <u>E</u>

#8   The minimum value is 10 and the maximum value is 29. The median is 19. Only A matches these criteria.

ANS: <u>A</u>

#9   The monthly median value is trending lower as the month increases. People born near end of the year tended to have lower lottery numbers than those born near the first of the year. No regression can be computed using only the medians.

ANS: <u>D</u>

#10   P-value is the measure of the probability of getting the sample result if the population mean is really what is hypothesized.

ANS: <u>A</u>

#11   Binomial distribution. Let X = number of people out of six randomly selected golfers who can hit a golf drive 300 yards.

$P(X \geq 4) =$

$\quad\quad P(X = 4) = {}_6C_4(0.35)^4(0.65)^2 = 0.0951$

$\quad + \quad P(X = 5) = {}_6C_5(0.35)^5(0.65)^1 = 0.0205$

$\quad + \quad P(X = 6) = {}_6C_6(0.35)^6(0.65)^0 = 0.0018 = 0.1174 \approx 0.117$

ANS: <u>B</u>

#12   289 <u>186</u> 9578 88<u>231</u> 33276 7<u>0997</u> 79936 56865 05859 90<u>106</u> 31595 0<u>1547</u> 85590
186, 231, 99, 106, 154

ANS: <u>D</u>

#13   Explanatory variable.

ANS: <u>B</u>

#14   Null hypothesis always includes the equals sign. Alternate hypothesis in this case is less than 48 oz.

ANS: <u>A</u>

#15   All expected values are 141. χ^2 value for this problem = 11.702 with 3 degrees of freedom. This places the P-value between 0.005 and 0.01.

ANS: <u>C</u>

#16   Since the confidence interval contains zero, there is no statistically significant difference between the two means.

ANS: <u>D</u>

#17   Median = 13. Q1 = 11. Q3 = 14. Minimum = 10 and Maximum = 15. Only box plot A matches all this information.

ANS: <u>A</u>

#18   Q1 is at the 19th position and equals 8. Q3 is at the 57th position and equals 12. InterQuartile Range = 12 - 8 = 4.

ANS: <u>B</u>

#19   Confidence intervals are used to estimate the population mean when only a sample of the data is known. The administrator has the entire population information.

ANS: <u>A</u>

#20 Any sampling distribution with a large enough sample size will take a normal distribution regardless of the underlying distribution.

ANS: <u>A</u>

#21 $\hat{p} = \dfrac{12}{400} = 0.03$, $p = 0.022$

$z = \dfrac{0.03 - 0.022}{\sqrt{\dfrac{(0.0220)(0.978)}{400}}} = 1.09$, $P(z > 1.09) = 0.14$

ANS: <u>A</u>

#22
A   True.   Confidence interval doesn't contain 0.
B   False.   There is no proof for this statement.
C   True.   Since there are 15 pairs, the df = 15 - 1 = 14.
D   True. $t_{Critical}$ will need to be adjusted based on degrees of freedom.
E   True.   SE MEAN is $\sigma_{\bar{x}}$.
Only false answer is B

ANS: <u>B</u>

#23 D is clearly the unique outlier. It has a large positive residual. If it were removed, the other four residuals would be randomly scattered above and below the x-axis.

ANS: <u>D</u>

#24 InvNormal(0.30) = -0.52 = z

$z = \dfrac{x - \mu}{\sigma} \Rightarrow -0.52 = \dfrac{x - 2000}{150} \Rightarrow x = (-0.52)(150) + 2000 = 1922$

ANS: <u>C</u>

#25 Course A: $z = \dfrac{80 - 76}{8} = 0.50$. Therefore, Course B needs to be played with a z-score of 0.50.

$0.50 = \dfrac{x - 80}{6} \Rightarrow x = 83$

ANS: <u>B</u>

#26 This is a matched pairs design testing the difference between the before and after times. Use a right tail, one sample t-test with degrees of freedom = 9 - 1 = 8. The $t_{Critical}$ value is 1.860.

ANS: <u>D</u>

#27 $Slope = r\dfrac{s_y}{s_x} \Rightarrow r = \dfrac{(Slope)(s_x)}{s_y} = \dfrac{(3.6)(2)}{8} = 0.9 \therefore r^2 = 0.81$.

ANS: <u>C</u>

#28 Probability $(X \geq 0.5)$ = area of triangle. A = ½bh = ½(0.5)(0.5) = 0.125 ≈ 0.13.

ANS: <u>E</u>

#29 A normal distribution has most of the values near the median and fewer values in the tails. In a box plot, the tails are longer than the middle boxes and the data should be symmetric. D is the best match for this even though it has an outlier.

ANS: <u>D</u>

#30 The mean or the expectation on each roll is $\$5(\frac{1}{3}) + (-2)(\frac{2}{3}) = \frac{1}{3}$ with standard deviation, $\sigma = 3.30$. For 30 turns, the expected winnings are $(30)\left(\frac{1}{3}\right) = \$10$. The standard deviation of the 30 turns $= \sqrt{30(3.30^2)} = 18.07$

$$P(X > 15) = P(z > \frac{15 - 10}{18.07}) = P(z > 0.2767)$$
$$P(z > 0.28) = 1 - 0.6103 = 0.3897$$

ANS: <u>E</u>

#31 P(Airport and Not SUV) = (0.70)(0.80) = 0.56
P(Downtown and Not SUV) = (0.30)(0.60) = 0.18
P(Not SUV) = 0.56 + 0.18 = 0.74 = 74%

ANS: <u>E</u>

#32 Robust procedures or intervals change very little when assumptions are not met.

ANS: <u>C</u>

#33 This is sampling without replacement. The probabilities change with each draw. P(3 Balls are not green) $= \left(\frac{7}{8}\right)\left(\frac{6}{7}\right)\left(\frac{5}{6}\right) = \frac{5}{8} = 0.625$.

ANS: <u>B</u>

#34 Fastest 10% have the "lowest" times. InvNormal(0.10) = -1.28
$$-1.28 = \frac{x - 43.5}{16.2} \Rightarrow x = (-1.28)(16.2) + 43.5 = 22.76 \text{ minutes.}$$

ANS: <u>A</u>

#35 Systematic. This is not a Simple Random Sample because each student doesn't have an equal chance of being selected. Also, not all subgroups of 223 have an equal chance of being selected.

ANS: <u>D</u>

#36 Because a treatment was imposed (one group got extra food), this is an experiment. However, since the guppies weren't assigned randomly to each side for food distribution, it is a poorly designed experiment.

ANS: <u>A</u>

#37 Choice I   True. The whiskers for the upper quartile for both groups are similar in length.
Choice II   True
Choice III   False. This shows a trend for a group but cannot be applied to an individual.
Choices I and II are true.

ANS: <u>C</u>

#38   Choice I   True
        Choice II   False
        Choice III   False
        Power is the measure of the ability of a hypothesis test to detect a difference between the actual mean and a hypothesized mean.

                                                                                        ANS: <u>A</u>

#39   Changing the order of the dependent and independent variables (response variable and explanatory variable) does not change the strength of the relationship. The value of r^2 is unchanged.

                                                                                        ANS: <u>C</u>

#40   Degrees of freedom = (rows - 1)(columns - 1) = (3 - 1)(9 - 1) = (2)(8) = 16.

                                                                                        ANS: <u>A</u>

Textbook Correlation for Practice Exam 4 – Multiple Choice

Question	Correct answer	Section	Question	Correct answer	Section
1	C	6.3	21	A	12.1
2	B	3.3	22	B	10.1
3	C	5.1	23	D	3.3
4	C	9.3	24	C	2.2
5	B	11.2	25	B	2.2
6	D	5.2	26	D	10.4
7	E	13.2	27	C	3.2
8	A	1.1	28	E	6.1
9	D	1.1	29	D	1.1
10	A	12.2	30	E	7.2
11	D	8.1	31	E	6.3
12	D	5.1	32	C	10.1
13	B	3.1	33	B	6.1
14	A	10.4	34	A	2.2
15	C	13.1	35	D	5.1
16	D	11.2	36	A	5.2
17	A	1.1	37	C	1.2
18	B	1.2	38	A	10.4
19	A	10.1	39	C	3.2
20	A	12.2	40	A	13.2

#1

(A)

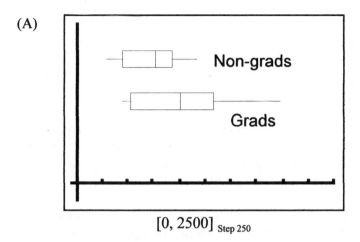

[0, 2500] Step 250

When comparing data sets, discuss center, spread, and shape.
The median for the non-graduates is lower than the median of the graduates. The minimum and Q1 values are reasonably close for both graphs. The Q3 value for the non-graduates is less than the median for the graduates. In fact, the maximum for the non-graduates is less than the Q3 value for graduates. The range of the non-graduate group is substantially smaller than the range for the graduate group. The data for the non-graduates is reasonably symmetric in shape with a slight skewness to the lower values. The shape for the graduates is decidedly asymmetric with skewness to the upper values.

(B) $\mu_{N+G} = \mu_N + \mu_G = \$700 + \$1000 = \$1700$

$\sigma_{N+G} = \sqrt{Var_N + Var_G} = \sqrt{125^2 + 250^2} = \$280$

(C)   Total salary $= (10)(\$700) + (15)(\$800) = \$19,000.$

$Mean = \dfrac{Total\,Salary}{Number\,of\,Employees} = \dfrac{\$19,000}{25} = \$760$

(D)   Total salary $= (15)(1000) + (5)(1050) = \$20,250.$

$Mean = \dfrac{Total\,Salary}{Number\,of\,Employees} = \dfrac{\$20,250}{20} = \$1,012.50$

(E)   Although the mean of both the non-graduates and the graduates increased, the overall mean decreased. This is an example of Simpson's Paradox. It happens because of unequal weighting of the new employees in relation to the number of workers already employed by the company. The non-graduates used to make up 40% of the company. With the new hires, the non-graduates now represent 56% of the company's workers. Since the mean of the non-graduates was lower than the company mean to begin with, the addition of new workers at a relatively low salary caused the overall company mean to drop.

(A)   There are 20 parcels on the north side and 20 on the south side. But the question does not deal with output based on location, only on the type of cutting. Number the parcels from 1 to 40 and use a random number table to select 20 plots of land.

Group 1: The first 20 will use selective cutting. Each year for 20 years, randomly select 10% of the mature trees for harvesting. Replant with a sufficient number of seedlings to insure the same number of trees when they mature. Report the number of trees harvested each year. Each parcel will have two full cuttings over the 20 years. Each year the equivalent of two full parcels will be harvested.

Group 2: The other 20 parcels will use clear cutting. Randomly allocate the twenty parcels into ten sets of two parcels each. Each year clear cut all the trees in the two specified parcels. Report the number of tree harvested each year. Replant with sufficient seedlings to guarantee the same number of mature trees. Each parcel will be clear cut twice during the 20 years.

(B)   The question asks which method produces the highest yield. No question regarding which side of the mountain was asked. For this reason a chi-square test is not required in this situation. Each year's harvest can be thought of as a random sample. There will be 20 samples for each type of cutting. A two-sample t-test of means can be used to compare the output for each type of harvesting.

Alternatively, each year's harvests can be paired. The test would then be a one sample   t-test on the difference between the two harvests each year.

#3

(A)   1986 corresponds to Year = 6. SALES = 15.5 + 2.86(6) = 32.66 billion dollars.

(B)   The residual for Year 6 is negative, which means the actual value is less than the predicted value. The value found in part (A) is an over-estimate.

(C) $Slope = r\dfrac{s_y}{s_x} \Rightarrow s_y = \dfrac{(Slope)(s_x)}{r} = \dfrac{(2.86)\sqrt{13.03}}{\sqrt{0.91}} = 10.82$

$\Rightarrow Var_y = 10.82^2 = 117.07$

(D)   For the first half of the time (1981-1987) the residual plot is nicely scattered about the x-axis. After 1987 the residuals have a wider variation and have a distinct pattern. It appears the model is good for the first part but is not good a model for the second half of the time of the study.

(E)   From the t-tables a t-statistic of 10.03 with 10 degrees of freedom has a P-value of less than 0.0005. This is very strong evidence against the null hypothesis in favor of the alternate hypothesis. It is extremely unlikely that the population this data is drawn from has a slope of zero.

#4

(A) The mean reading score cannot be determined since the two of the values are categorical, below 5.5 and 10+. It is impossible to add these to the values of the other reading scores.

(B) Claim:

  The new reading program has improved the reading ability of the students.

  Counterclaim:

  The new reading program has not improved the reading ability of the students.

  Hypotheses:

  H_0: There is no relationship between reading level and date of the test.

  H_a: There is a relationship between reading level and date of the test.

  Appropriate test:

  χ^2 test for independence.

  Level of significance:

  Let $\alpha = 0.05$ (stated in problem).

  χ^2 requires values, not percentages, so the table needs to be rewritten with values for each cell. (Expected values are in parentheses.)

Reading Level (grade level)	September 2001	March 2002
Below 5.5	30 (24)	18 (24)
6	63 (44)	25 (44)
7	47 (44)	41 (44)
8	137 (127)	117 (127)
9	114 (121)	128 (121)
10+	33 (64)	95 (64)
Totals	424	424

Conditions and Assumptions for a Two-way Test for Independence:

  1. The data come from an SRS. Stated in the problem.
  2. The data are independent. It is unlikely that one student's score influenced another student's score.
  3. All cells have expected values of at least 24, which is greater than 5.

All conditions and assumptions have been met to proceed with a χ^2 test.

$$\chi^2 = \sum \frac{(observed - expected)^2}{expected} = 52.2 \text{ (By calculator).}$$

P-value ≈ 0.

Decision:

   Reject H_0

From the data table it is clear that March values at the lower end of the reading levels are lower than expected and the values at the upper end of the reading levels are higher than expected.

Conclusion:

   There is sufficient data to support the claim the new method of teaching reading has improved the reading scores of students.

(A)   Back-to-back stem plot

Class 1		Class 2
4	0	
4	1	8
64421	2	59
0	3	0
600	4	6
	5	457
7651	6	015
30	7	
752	8	34699
88540	9	28
0	10	00

$1|8 = 18$

The scores from Class 1 tend to be lower than the scores from Class 2. The range for Class 1 is a little larger than the range for Class 2. Neither class has an outlier. The samples are not large enough to establish a normal distribution. However, at the same time, there is no evidence to state the data is non-normal in shape.

(B)
Claim:

The proportion of the population of students that passed the exam is the same in both classes.
Counterclaim:

The proportion of the population students that passed the exam is not the same for both classes.

$H_0: p_1 = p_2$

$H_a: p_1 \neq p_2$ Where p_1 is the proportion of students who passed in Class 1 and p_2 is the proportion of students who passed in Class 2.

Let $\alpha = 0.05$

$$\hat{p}_1 = \frac{11}{26} = 0.423, \quad \hat{p}_2 = \frac{9}{20} = 0.45$$

Assumptions and conditions:

1. Data is from a random sample. Stated in problem.
2. Data is independent. There is no reason to believe that one score affected another score.
3. Data is not extremely non-normal in shape. Histogram from part (A) verifies this.

$$\text{Also } n_1 \hat{p}_1 = 26 \times \frac{11}{26} = 11, \text{ and } n_1(1 - \hat{p}_1) = 26 \times \frac{15}{26} = 15$$

$$\text{and } n_2 \hat{p}_2 = 20 \times \frac{9}{20} = 9, \text{ and } n_2(1 - \hat{p}_2) = 20 \times \frac{11}{20} = 11$$

All values are greater than 5.

Since all assumptions and conditions are met. We can proceed.

Use a two-proportion z-test.

$$z = \frac{0.423 - 0.45}{\sqrt{\frac{(0.423)(0.577)}{26} + \frac{(0.45)(0.55)}{20}}} = \pm 0.18$$

$P(z \le -0.18 \text{ or } z \ge 0.18) = 0.8572$ (by calculator P = 0.8551)

Decision:

       Fail to reject H_0

Conclusion:

       There is not sufficient evidence to reject the claim that the proportion of students that passed the exam is the same in both classes. It is likely that both classes had a similar rate of passing.

#6

(A)　　Group A　　Starting with the first line of the random number table, use one-digit numbers.

| 08<u>421</u> | 44753 | 77377 | 28744 | 75592 | 08563 | 79140 | 92454 | 14592 | 98481 | 15373 |

Schools chosen are 4, 2, and 1.

　　Group B　　Starting with the second line of the random number table, use two-digit numbers.

| 12<u>109</u> | 47<u>806</u> | 11421 | 66812 | 53<u>045</u> | 66831 | 68<u>098</u> | 40772 | 21558 | 47781 | 33586 |

Schools chosen are 10, 6, 4, and 9.

　　Group C　　Starting with the third line of the random number table, use two-digit numbers.

| 79<u>177</u> | 06928 | 70708 | 23822 | <u>13048</u> | 12633 | <u>19687</u> | 05233 | <u>16853</u> | 14374 | 40011 |

Schools chosen are 17, 13, 4, 19, and 16.

Compliance

Group A		Group B		Group C	
School #	Compliance	School #	Compliance	School #	Compliance
4	Yes	10	No	17	Yes
2	No	6	Yes	13	Yes
1	Yes	4	No	4	Yes
		9	Yes	19	No
				16	Yes
2 of 3 are in compliance		2 of 4 are in compliance		4 of 5 are in compliance	

(B)　　In the sample 8 of 12 schools are in compliance or 0.67.

(C)　　True proportion of schools in compliance is 16 of 35 or 0.46.

(D)　　A confidence interval requires a random sample be drawn from the entire population. These data were taken by a stratified sample which violates this assumption.